INTRODUCING TEACHERS

— + —

ADMINISTRATORS

TO

THE

NGSS

A
PROFESSIONAL
DEVELOPMENT
FACILITATOR'S
GUIDE

INTRODUCING
TEACHERS + ADMINISTRATORS
TO THE
NGSS

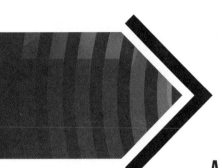

A PROFESSIONAL DEVELOPMENT
FACILITATOR'S GUIDE

Eric Brunsell

Deb M. Kneser

Kevin J. Niemi

NSTApress

National Science Teachers Association

Arlington, Virginia

National Science Teachers Association

Claire Reinburg, Director
Wendy Rubin, Managing Editor
Andrew Cooke, Senior Editor
Amanda O'Brien, Associate Editor
Amy America, Book Acquisitions Coordinator

ART AND DESIGN
Will Thomas Jr., Director
Joe Butera, Senior Graphic Designer, cover and
    interior design

PRINTING AND PRODUCTION
Catherine Lorrain, Director

NATIONAL SCIENCE TEACHERS ASSOCIATION
David L. Evans, Executive Director
David Beacom, Publisher

1840 Wilson Blvd., Arlington, VA 22201
www.nsta.org/store
For customer service inquiries, please call 800-277-5300.

Library of Congress Cataloging-in-Publication Data.
Brunsell, Eric, author.
  Introducing teachers and administrators to the NGSS : a professional development facilitator's guide /
Eric Brunsell, Deb M. Kneser, Kevin J. Niemi.
     pages cm
  Includes bibliographical references.
   ISBN 978-1-938946-18-9 -- ISBN 978-1-938946-60-8 (e-book) 1. Science--Study and teaching--Standards-
-United States. 2. Educational accountability--United States. I. Kneser, Deb M., 1955- author. II. Niemi,
Kevin J., 1956- author. III. Title.
  Q183.3.A1B78 2014
  507.1'273--dc23
                              2014010659
Cataloging-in-Publication Data for the e-book are available from the Library of Congress.

CONTENTS

CONTENTS

CONTENTS

ABOUT THE AUTHORS

ERIC BRUNSELL is an associate professor of science education in the Department of Curriculum and Instruction and coordinator of the Center for Excellence in Teaching and Learning at the University of Wisconsin-Oshkosh in Oshkosh, Wisconsin. Eric earned his EdD in Curriculum and Instruction with an emphasis in Science Education from Montana State University. He is a former high school science teacher and has served as NSTA District 12 Director. Eric has provided professional development sessions and presentations throughout the United States and internationally.

DEB M. KNESER is an assistant professor in the School of Education and the chair of the Institute of Professional Development at Marian University, Fond du Lac, Wisconsin. Deb earned her Ph.D. in curriculum leadership from Marian University. She was previously with the Cooperative Educational Service Agency 6 in Oshkosh, Wisconsin, serving as an educational consultant in curriculum, assessment, and instruction. Deb is a former elementary classroom teacher with a National Board Certification. Deb has extensive experience with providing professional development sessions and presentations on topics ranging from curriculum mapping to teacher effectiveness.

KEVIN J. NIEMI is director of the outreach group in the Institute for Biology Education at the University of Wisconsin-Madison. Kevin received his Ph.D. in plant physiology from the University of Minnesota, taught in the biology department at Grinnell College, and served as the first education coordinator for Olbrich Botanical Gardens in Madison, Wisconsin. Kevin has provided professional development sessions and presentations in science education in Wisconsin and Thailand.

CONTRIBUTORS

The following educators contributed to the development of many activities in this book. We would like to thank them for their help, and we consider them coauthors of the activities they assisted with (indicated in the table of contents).

Mark Bazata
Instructional support teacher,
high school,
Oshkosh Area School District,
Oshkosh, Wisconsin

Reynee Kachur
Department of Biology and
Microbiology,
University of Wisconsin-Oshkosh,
Oshkosh, Wisconsin

Emily Miller
NGSS Writing Team
Elementary bilingual resource teacher,
Madison Metropolitan School District,
Madison, Wisconsin

Chad Janowski
Chair, Science Department, high school,
Shawano School District,
Shawano, Wisconsin

Amy Parrott
Mathematics Department,
University of Wisconsin-Oshkosh,
Oshkosh, Wisconsin

Stacey N. Skoning
Chair, Department of Special Education,
University of Wisconsin-Oshkosh,
Oshkosh, Wisconsin

A Letter From David L. Evans, NSTA Executive Director

Pick up a newspaper and you'll see stories related to legalizing marijuana, internet privacy concerns, the gluten-free diet, and hydraulic fracturing (or "fracking"). Strike that. Who reads an actual *paper* anymore? Now we get our news reports in the palms of our hands with pocket-sized devices that also enable us to call friends, get weather and traffic reports, take pictures, listen to music, turn off the lights at home, and so on.

Every day, we make decisions that require a fairly high level of scientific literacy. *Should I buy antibacterial soap? Should I vaccinate my children? Should I buy organic fruits and vegetables?* Everywhere you turn, you see further evidence that we live in an increasingly technological world, a world supported by jobs and industries we couldn't imagine even two decades ago. *App developer? Bitcoin?*

We need to prepare our children not just to live but to thrive in a world we can't foresee. They *all* need high-quality science instruction. They need to understand how to think and behave like scientists and engineers. The *Next Generation Science Standards (NGSS)* outline an approach to science education that helps students understand

important concepts within the context of real-world skills and applications, that helps them draw connections between and among science, engineering, math, and English language arts (ELA). These standards are based on decades of sound research, as summarized in *A Framework for K–12 Science Education*.

NSTA was a partner in the development of the *NGSS* and is committed to helping administrators and teachers better understand and implement the standards in their schools and classrooms. The shift in instruction and thinking can be overwhelming. NSTA encourages a thoughtful approach to implementation—with collaboration with colleagues being key—and NSTA offers an ever-growing collection of resources to help educators every step of the way.

Start by visiting the NGSS@NSTA Hub at *www.nsta.org/ngss*, the gateway to the full spectrum of NSTA's *NGSS*-related products and services. Here you'll find a user-friendly presentation of the *NGSS* performance expectations with related practices, crosscutting concepts, and core ideas. View the full standard page or isolate specific performance expectations with their corresponding dimensions. In addition, you'll find resources vetted by a group of NSTA curators and tagged to particular performance expectations. Other tools include streamlined charts of the dimensions, as well as of the relationships among science, math, and ELA practices. These resources are particularly useful when leading a team of teachers through a training workshop.

The NGSS@NSTA Hub will continue to evolve, as we add functionality and new content. However, it will always be NSTA's central spot for information and resources around the standards, including the latest news on adoption and assessment. And it's where we'll showcase upcoming events and opportunities such as special conference sessions and professional development institutes, virtual conferences, and online short courses.

In addition, administrators will appreciate NSTA's *NGSS* publications—especially *The NSTA Reader's Guide to A Framework for K–12 Science Education* and *The NSTA Reader's Guide to the Next Generation Science Standards.* Taken together, these slim and practical volumes will help you introduce teams of teachers to the three dimensions and new standards, then help you coach them through the planning and implementation phases. Both books are available in print or digital formats. In addition to this book, another valuable publication is Rodger Bybee's *Translating the* NGSS *for Classroom Instruction,* which helps bridge the gap between standards and practice,

and the elementary-level *Science for the Next Generation* (Banko et al.), which approaches the new standards via the popular and effective 5E Model. In addition, browse the NSTA Learning Center for NSTA's full collection of journal articles, including several special series on the *NGSS,* as these pieces provide excellent foundations for working group discussions.

NSTA has also developed an archive of free web seminars covering each of the science and engineering practices, disciplinary core ideas, and crosscutting concepts in detail. Share these with colleagues, and encourage your teachers to gain familiarity and confidence with the idea of using the practices to teach the content. Future web seminars will delve into grade-specific standards.

During this extraordinary time in science education, the challenges and stakes are great, but so are the opportunities. NSTA's goal continues to be to support excellent and innovative science instruction for *all* students. To achieve that goal, we are committed to helping science teachers and administrators by developing the tools you need to successfully understand and implement the *Next Generation Science Standards.*

David L. Evans

David L. Evans
NSTA Executive Director

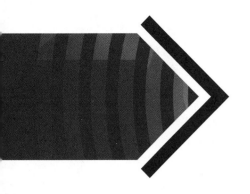

1

Introduction

Science learning in school leads to citizens with the confidence, ability, and inclination to continue learning about issues, scientific and otherwise, that affect their lives and communities.

—*A Framework for K–12 Science Education* (NRC 2012, pp. 286–287)

Throughout our careers as educators, we (the authors) have each attended dozens of professional conferences as participants, presenters, and even as vendors. We leave these conferences enriched and energized by the sharing of teaching resources, strategies, and lessons learned. It is always exciting to engage with like-minded professionals as we explore our craft. However, the 2013 NSTA National Conference on Science Education was unlike any other. The excitement had been building for months, and the energy was palpable—not just on San Antonio's Riverwalk! As thousands of science educators from across the country began converging on the conference, the *Next Generation Science Standards* (*NGSS*; NGSS Lead States 2013) were released. It quickly became apparent that this effort surpassed expectations and held the potential to transform science learning for students throughout the United States.

For us, the release of the *NGSS* marked a major turning point in our professional efforts. Over the past two years, we have been working with science education leaders from across Wisconsin in a loosely coordinated grassroots effort to build understanding of *A Framework for K–12 Science Education* (*Framework*; NRC 2012) and the development process for the *NGSS*. From the initial draft and publication of the *Framework* to the release and review of two *NGSS* drafts, we witnessed a growing excitement for and anticipation of the final publication of the standards.

As we transitioned to facilitating professional development to help teachers and administrators begin implementing these new standards, our understanding of the complexity of the *NGSS* deepened. In our workshops, participants are consistently positive about the potential impact. However, at times there is also a sense of being overwhelmed; the introduction of the *NGSS* comes in the midst of massive changes in the educational landscape, including implementation of the *Common Core State Standards* (NGAC and CCSSO 2010), in English language arts and mathematics; disciplinary literacy and Response to Intervention initiatives; the transition to new accountability systems; and dramatic changes to teacher evaluation processes. Effective and strategic professional development is critical for ensuring alignment across these disparate efforts, or we risk losing the potential of *NGSS* in a haze of initiative fatigue. This book provides tips, suggestions, resources, and a series of activities that can inform initial professional development experiences around the *NGSS*.

Chapter 2 discusses how to use this book to plan professional development. It includes a description of two "strands" of work that schools need to engage in during the implementation process, provides guideposts for effective professional development, and identifies a series of planning steps. This chapter also includes a table that summarizes all the activities in the book. Chapter 3 identifies a series of challenging questions and issues that arise frequently during professional development related to *NGSS*. Chapter 4 provides suggestions for administrators as they consider how to lead and support implementation of the *NGSS*.

Chapters 5–9 describe 24 activities that can be used as starting points for examining the curricular and pedagogical changes necessary for implementation of the *NGSS*. Chapter 5 focuses on introducing the vocabulary and structure of the *NGSS* and deepening understanding of the conceptual shifts that guided their development. In Chapter 6, we provide a series of activities to help science teachers understand the three dimensions of the *Framework*—science and engineering practices, crosscutting concepts, and disciplinary core ideas—and how they are integrated in the *NGSS*. Activities in this chapter also build connections between the science and engineering practices, science inquiry, and engineering design. Chapter 7 includes activities that explore the *NGSS* Appendix D, "All Standards, All Students" and the case studies included with that appendix. These classroom case studies describe instructional approaches for groups of students often challenged by traditional science teaching. Chapter 8 focuses on beginning discussions related to curricular decisions including course mapping, designing essential questions and performance assessments, and using the *NGSS* to plan units of instruction. In Chapter 9, activities examine the connections between *NGSS* and the *Common Core State Standards*.

Realizing the vision of *A Framework for K–12 Science Education* and the *Next Generation Science Standards* will require professional development of an unprecedented scale. Our hope is that this collection of activities will be helpful as you lead educators through the implementation process.

2

Facilitating Professional Development Using This Book

Initial professional development efforts related to the *Next Generation Science Standards* (*NGSS*; NGSS Lead States 2013) have understandably focused on building awareness, but we must quickly transition to providing sustained experiences that give teachers opportunities to construct pedagogical and content knowledge related to *NGSS* through collaboration, classroom enactment, and reflection. There is no "easy button" for implementing the *NGSS*; implementation is not something that can be done in a weeklong institute or even over a few months in the summer. The activities in this book are intended to provide starting points for helping administrators and teachers of science understand the complexity of effective science teaching and the *NGSS*. We need to stress this point: No single book can provide everything needed for implementation. In this chapter, we outline two strands of work that we think districts will need to undertake for *NGSS* implementation, briefly summarize research on effective professional development, and provide a series of steps for using this book to plan a professional development experience.

Implementation Efforts

In our work with teachers and administrators, we see two distinct strands of professional development that educators need to engage in to successfully implement *NGSS*. In addition to these strands, educators will need to frequently communicate with the school community (including parents) and stakeholders beyond the school about changes to curriculum and instructional approaches.

Strand One: Curricular Decisions

When we first started doing professional development related to *NGSS*, we quickly realized that the first questions from both teachers and administrators focused on curricular considerations: How will *NGSS* change the sequence of courses, and how will it change the content? We found that it was often difficult to delve into pedagogical activities and discussions without giving some attention to (1) the sequence of topics and courses in elementary, middle, and high school; and (2) the content included—and excluded—from specific courses. The *NGSS* define the content by grade level in elementary school, but provide tremendous flexibility for curriculum in grades 6–12. Choosing the sequence of content is a difficult decision that requires strong leadership. The second point, inclusion and exclusion of content, is particularly important. If administrators and teachers treat *NGSS* as a checklist and simply add more content to fill gaps, we will be doing a disservice to the students. Instead, we need to design courses that allow students to engage deeply with a focused set of foundational concepts. Activities 14–17 in Chapter 8 (see Table 2.1, pp. 8–13 summary table of activities) help begin the discussions related to designing a scope and sequence aligned with *NGSS*. We encourage you to begin this process early in your implementation plan.

Strand Two: Pedagogical Development

The expectations defined in *NGSS* will require many educators to teach in a new way. Although inquiry was called out as a cornerstone of good science instruction by the original *National Science Education Standards* more than 15 years ago, implementation of an inquiry approach is far from universal in our classrooms. The *NGSS* provide a much clearer description of the work of scientists and engineers by defining a series of eight science and engineering practices. In addition, the *NGSS* emphasize that these practices cannot be taught isolated from content. Teachers of science will need to develop an understanding of each practice and determine strategies for building student capacity with these practices throughout the school year.

Similarly, the *NGSS* define seven crosscutting concepts. These crosscutting concepts can be understood as big ideas, or lenses, that scientists and engineers use to help them understand the natural world. For example, energy, as a crosscutting concept, can help scientists understand interactions, ecosystems, and chemical reactions. Just like the practices, crosscutting concepts must be developed within the context of content through repeated exposure.

Many of the activities in Chapters 6 and 7 can be used to deepen teachers' understanding of the science and engineering practices, crosscutting concepts, and disciplinary core ideas. However, these activities are only a start to that process. Mastery will only come through dedicated work over time.

Perhaps the most profound change in *NGSS* is the use of performance expectations to guide assessment. The *NGSS* ask educators to assess students differently than in previous standards: If students are expected to learn science by integrating the three dimensions, then they should also be assessed by their ability to perform using those three dimensions. Activity 19 in Chapter 8 begins the discussion of assessment, while Activities 18 and 20 help teachers think through the process of integrating the three dimensions while planning instruction.

We feel that it is important for schools to focus on both strands concurrently. Teacher teams

FIGURE 2.1

Four guideposts for designing effective professional development

Professional development should be COHERENT.	Professional development should be INTEGRATED.
• Is your experience connected to school or district goals? • Are you engaging teams, including teachers and administrators? • Is your experience connected to a comprehensive *NGSS* implementation plan? • Is your experience compatible with other professional development initiatives?	• Can your experience take advantage of multiple delivery formats (face-to-face, online, hybrid)? • Have you considered the structure you will use (workshop/institute, study groups, embedded in a professional learning community, action research, etc.)?
Professional development should be ENGAGED.	**Professional development should be PRACTICAL.**
• Do you provide multiple opportunities for active learning? • Does your experience embrace shared expertise and collaboration of participants? • Do you provide opportunities for participant meaning-making, debriefing, and reflection?	• Are your learning experiences relevant to participants' classrooms or job responsibilities (e.g., curriculum planning, experiencing model activities)? • Do you make concrete connections between theory and practice? • Do you encourage enactment in the classroom? Is it possible to include examples of student work? • Is participant time being used productively in goal-oriented activities?

should be engaged in scope and sequence discussions while also deepening their understanding of the pedagogical changes described in *NGSS*.

Effective Professional Development

A growing body of research has identified professional development practices that lead to continued growth and change in instructional practice. Effective professional development should be

measured against its capacity to equip teachers as "shapers, promoters, and well-informed critics of reform" (Little 1993, p. 130). Fogarty and Pete (2010) describe a "syllabus of seven" protocols or elements that anchor professional learning experiences for lasting impact. These elements include professional learning that is sustained, job embedded, collegial, integrative (multiformat), practical, and results oriented. Similarly, Hunzicker (2011) has provided a research-based checklist that can be used as a planning tool for designing effective professional development. Hunzicker's checklist consists of five categories: supportive of the needs of individuals and school or district goals, job-embedded, "instructional focus," collaborative, and ongoing. Figure 2.1 (p. 5) combines these characteristics of effective professional development into four guideposts to assist you as you plan professional development for implementation of the *NGSS*.

Planning a Professional Development Experience Using This Book

Our intention in this book was not to provide a "one size fits all" series of professional development activities that can be done sequentially. Instead, we hope to provide a collection of activities that can be packaged in different ways to meet a variety of needs. As a result, your first step in designing any experience using this book is to clearly define your audience and purpose. This will allow you to select appropriate activities to use within a professional development plan consistent with the four guideposts: coherent, integrated, engaged, and practical. Table 2.1 (pp. 8–13) provides a summary of each activity. In each of the three examples below, a variety of activities from this book are used to meet the needs of different audiences.

1. *You have one hour to introduce NGSS to an audience with very little familiarity.* Since your goal is to introduce *NGSS*, you will want to focus on activities in Chapter 5. To allow participants to engage immediately with the standards, you use Activity 1 for the first 20 minutes. After participants have had a chance to examine and discuss a standards progression, you use Activity 3 to help

participants understand the structure and organization of the *NGSS*. At the end of the session, you use exit slips to collect feedback from participants.

2. *You have three hours with a mixed group of administrators and educators.* Since you are working with a group that includes administrators, you decide to focus on the impact of *NGSS* on the scope and sequence of science curriculum. You also know that many in attendance are not familiar with the *NGSS*. Therefore, you begin the session with Activity 4 to help participants understand the structure of *NGSS* and the conceptual shifts made in the standards. You then split the group into elementary, middle, and high school teams. You use Activity 15 with the elementary group and Activity 17 with the middle and high school groups so that they can all be engaged in exploring possible course sequences.

3. *You have monthly meetings with your science team.* You work closely with these teachers on a regular basis and know that they already have an introductory understanding of *NGSS*. Therefore, you begin the year with Activities 8, 5, and 6 to have teachers self-assess (Activity 8) their understanding of the practices and to begin discussions about the role of science and engineering practices (Activities 5 and 6). You then use additional resources (such as NSTA's *NGSS* webinars) to continue to focus on explanations and argumentation (science and engineering practices) with classroom implementation. After working on this for a few months, you can use Activity 19 to start collaborating on developing performance assessments.

When planning the logistics for your professional development experience, we suggest the following steps:

1. Become familiar with the activities you will use. Work through the activity as a participant and as a facilitator. Take notes about specific transitions that need to be made and points that you want to emphasize. Make sure that the facilities are suitable for the activities you have selected.

2. Think like an assessor. When possible, reach out to participants in advance to identify their needs and goals for attending. Also, determine information that you can collect from participants to improve and guide future sessions. Many of the activities in this book include specific prompts or handouts you can use to collect data. We also often use an exit slip that includes the following prompts:

    a. What worked well today?

    b. What did not work well?

    c. I have the following questions or could use additional support with the following.

    d. Please provide your name if you would like us to follow-up with you on specific questions.

    In addition to exit slips, we employ a "parking lot" for questions and issues that arise during the session. We simply post a piece of labeled chart paper on the wall. If questions arise that we cannot answer, or issues surface that we feel will distract from the flow of the session, we add them to the parking lot for future follow-up.

If you do collect data from participants or use a parking lot, it is critical that you follow through. If you are meeting with the participants again, begin by summarizing the data and describing how you will act on the questions or issues. If you are not meeting again, you can use a follow-up e-mail.

3. Plan a detailed schedule for the session. Include a timeline and the actions and activities that you will use to activate prior knowledge, build knowledge, and debrief. Allocating time at the end of a session to debrief and reflect is critical. If time is running short, make sure that you stop other activities early enough to provide participants time to reflect. Your schedule should also include notes on how you plan to transition from one activity to the next. Finally, include notes to yourself about areas that you can cut if you are running out of time and areas that can be stretched if you are progressing faster than expected. Remember, show that you respect your participants' time commitment by starting on time and ending on time ... or early!

4. Prepare your presentation and assemble materials. As you read through the activities, determine if you need to create a presentation. However, keep these short. Too much text on a slide distracts (and bores) participants. Dedicate as much time as possible to active learning. In addition, many of the activities require access to the *NGSS*. Determine if you can provide internet access or hard copies. If participants are expected to bring their own copy, make sure that they are aware in advance. You should also assemble the materials needed for your

activities and organize them so that they are easy to pass out during the session.

5. On the day of the session, prepare the room. If possible, arrange the room to encourage participants to discuss with each other. Make sure that your materials are easily available. In addition, do a dry run through the session and check any technology that you plan on using.

6. Finally, after the session, reflect on your performance, review any of the data you collected, and begin planning your next steps.

TABLE 2.1

Activity summaries

Activity	Title	Estimated time (min)	Topic	Summary
1	**Examining the Standards**	30–40	Introduction to *NGSS* progressions	Participants explore one strand (either organized by topic or disciplinary core idea) of the standards from early elementary through high school.
2	**NGSS Vocabulary**	45	Development and structure of *NGSS*	This activity provides an overview of a standards page, and is one of three introductory activities included in this book.
3	**The Structure of NGSS**	55	Organization and structure of an *NGSS* page	This is the second of three activities in this book used to introduce educators to the structure of the *NGSS*. In this activity, participants use one of the introductory sections of the *NGSS* to determine their own definitions for important elements of *NGSS* structure (performance expectations, disciplinary core ideas, foundation boxes, and connections boxes).

Table 2.1 (*continued*)

Activity	Title	Estimated time (min)	Topic	Summary
4	*NGSS* **Conceptual Shifts**	55	Conceptual shifts desired for teaching with *NGSS*	This is the third activity used to introduce the development and structure of the *NGSS*. This activity focuses on the six conceptual shifts that demonstrate how *NGSS* is different from previous standards documents.
5	**Integrating the Three Dimensions**	180–220	Three dimensions of *NGSS*	Participants experience a model activity as they develop an understanding of the integration of the three dimensions of the *Framework* and the *NGSS*.
6	**Science Inquiry and the Practices of Science and Engineering**	60–90	Science and engineering practices	Participants deepen their understanding of science inquiry by exploring the connections between the *Framework*'s spheres of activity for scientists and engineers and the science and engineering practices of the *NGSS*.
7	**Engineering Design and the Practices of Science and Engineering**	110–120	Engineering design	Participants engage in a model activity as they develop an understanding of the engineering design process (as defined in the *NGSS*) and the science and engineering practices.
8	**Science and Engineering Practices Self-Assessment**	30–40	Practices check-up	Participants self-assess their understanding of the science and engineering practices. After the self-assessment, group data is collected related to the practices that participants are most and least comfortable with.

Table 2.1 (*continued*)

Activity	Title	Estimated time (min)	Topic	Summary
9	**Exploring Crosscutting Concepts**	60–80	Crosscutting concepts	Participants explore the crosscutting concepts in *NGSS*. They begin by organizing statements from the crosscutting concepts matrix into grade band progressions, are then introduced to the importance of crosscutting concepts, and finally identify key crosscutting concepts that are strongly connected to what they currently teach.
10	**Integrating the Nature of Science**	90	Nature of science	Participants engage in a model activity as they learn about how the nature of science is reflected in the *NGSS*. They consider their understanding of the nature of science, understand nature of science expectations in *NGSS*, and create draft anchor charts to develop student understanding of the nature of science.
11	**All Standards, All Students: Integrating Content, Practices, and Crosscutting Concepts**	60–90	*NGSS* are for all students	This is the first in a series of three activities related to Appendix D, "All Standards, All Students" from the *NGSS*. First, teachers become part of an "expert group" as they examine one case study for examples as to how the teacher in the vignette integrates disciplinary core ideas, science and engineering practices, and crosscutting concepts. Next, teachers form "jigsaw groups" to look for similarities across vignettes.

Table 2.1 (*continued*)

Activity	Title	Estimated time (min)	Topic	Summary
12	**All Standards, All Students and Universal Design for Learning**	70	Universal design approach for all students learning	Participants study the principles of Universal Design for Learning, identify strategies from vignettes that exemplify these principles, and reflect on actions that they can take in their own teaching.
13	**All Standards, All Students: A Strategy Matrix**	70	Science learning for nondominant groups	Participants learn about the key features of effective strategies from the research literature on science learning for nondominant groups, identify examples of strategies from the vignettes that fit with these features, and reflect on actions that they can take in their teaching.
14	**Visioning and Values**	50–60	Articulate the ideal science learning vision	In this activity, teams of teachers identify what values and aspirations they have for science at their school. Participants then write an "elevator speech" that summarizes these values.
15	**Course Mapping at the Elementary Level**	85	Topical analysis of NGSS, grades K–5	This activity allows participants to explore that scaffolding and see where topics fit within grades K–5. Participants will gain a better understanding of content that is being taught at the various grade levels.

Table 2.1 *(continued)*

Activity	Title	Estimated time (min)	Topic	Summary
16	**Plus, Minus, Delta (Grades 6–12)**	45	Topical analysis of *NGSS*, grades 6–12	Participants examine the *NGSS* standards for their grade band (6–8, 9–12) specific to the topics that they are currently teaching. Participants determine what current topics are well aligned with *NGSS*, which might be possible to de-emphasize or cut, and identify what changes may be necessary.
17	**Course Mapping for Middle and High School**	60	Course sequencing with *NGSS*, grades 6–12	This activity provides a structure for beginning to think through the process of clustering performance expectations into a course sequence at the middle and high school level.
18	**Essential Questions and Crosscutting Concepts**	45–50	Big ideas, essential questions, and crosscutting concepts with *NGSS*	This activity is designed to demonstrate one process for designing essential questions or "what do we want to learn" using the *NGSS* crosscutting concepts. Participants will be involved in a short discussion on the "why we need to do this" and "how we do it." Next, participants are asked to try and write several essential questions on their own to demonstrate their understanding of this step in planning a unit.

Table 2.1 (*continued*)

Activity	Title	Estimated time (min)	Topic	Summary
19	**Developing Performance Assessments**	75	Assessing and the *NGSS*	This activity allows participants to explore developing performance assessment tasks using the performance expectations from the *NGSS*. The focus of this activity is to help participants understand how to use the performance expectations to begin to design performance assessments.
20	**From Standards to Units**	80	*NGSS* and unit planning	This is an introductory activity that allows participants to explore the connections between *NGSS* and unit plans.
21	***NGSS* and the *CCSS* Mathematics**	140	*NGSS* and the *CCSS*, in mathematics	This activity introduces educators to the connections between the *NGSS* and the *CCSS*, in mathematics.
22	**Connecting *NGSS* and *CCSS* ELA**	50	*NGSS* and *CCSS*, in English language arts	This activity introduces educators to the connections between the *NGSS* and the *CCSS*, in English language arts.
23	***NGSS* and *CCSS* ELA: Connecting Through the Practices**	45–50	*CCSS*, in English language arts through the practices of *NGSS*	Participants examine science articles and provide evidence that the author uses to support his or her position. This activity demonstrates the relationship between the science and engineering practices and the *CCSS ELA*.
24	***NGSS* and *CCSS* ELA: Disciplinary Literacy**	60	*CCSS ELA*, disciplinary literacy in science, and *NGSS*	This activity introduces participants to disciplinary literacy and the connections to *NGSS*. Participants are given the opportunity to explore different styles of text and reading.

3

STICKING POINTS

During April and May of 2013, we facilitated workshops, inservice meetings, and an online course focused on the *Next Generation Science Standards* (*NGSS*; NGSS Lead States 2013) for more than 600 science educators. Although the majority of these educators were from Wisconsin, educators from 20 states participated in the online course. Through discussions, reflections, and workshop evaluations, we were able to identify a number of issues—sticking points—that tended to generate anxiety and bog down discussions if not addressed candidly. For many of these questions, answers simply are not known. In these cases, you should avoid the urge to speculate and start a "parking lot" for issues that you can follow up on later.

Is It *Common Core*?

K–12 education is in a turbulent transitional time, without a clear vision for how the myriad policy changes fit together. Teachers and administrators are suffering from initiative fatigue; elementary teachers in particular are being pulled in more directions than ever before. Many are in the midst of implementing the *Common Core State Standards* (*CCSS*; NGAC and CCSSO 2010), in mathematics and English language arts (ELA), and the thought of making drastic changes to the way they teach science can be overwhelming. However, elementary teachers often quickly see the power of the *NGSS* and are excited by the clarity of the grade-level structure.

At the middle and high school level, many science teachers are beginning to realize that the *CCSS ELA* also impact science. Although some schools have focused on disciplinary literacy for years, the *CCSS* provide a renewed sense of urgency.

Understanding that *NGSS* and *CCSS* are two different national (but not federal) efforts can cause confusion for many teachers. It is important to help participants sort through the different "brands" and the differing developmental processes. Additionally, it is important to build an understanding of *NGSS* as content standards and *CCSS ELA* (for science and technical fields) as literacy standards that are complementary. Implementing *NGSS* with fidelity will also result in more effective implementation of disciplinary literacy and the *CCSS ELA* literacy standards for science and technical fields.

How Will This Be Assessed?

This sticking point is closely related to the conflation of *NGSS* and *CCSS*. State-level accountability systems are changing to match the *CCSS*. Many states are using computer-based assessment systems as part of the Partnership for Assessment of Readiness for College and Careers (PARCC) or Smarter Balanced assessment consortia. However, these assessments do not assess the performance expectations in the *NGSS*. States adopting *NGSS* still need to assess science, but it is not currently clear how (or when) those assessments will change. Questions about assessment often become contentious because scores can be tied into school and teacher evaluations.

Policy discussions related to assessment are in their early stages. These discussions will be greatly impacted by the reauthorization (or lack thereof) of the federal Elementary and Secondary Education Act (No Child Left Behind). As a professional development facilitator, it is important that you stay informed about these policy discussions. You should be candid about the state of these issues and resist the urge to speculate.

It is important to convey that implementation of *NGSS* involves increasing the depth of students' engagement with foundational content, providing experiences that reflect the culture of science through the practices, and developing connections between science concepts through crosscutting concepts. Educational research has shown that this approach has a positive impact on standardized test scores. Additionally, implementation of *NGSS* complements disciplinary literacy efforts and should positively impact achievement on assessments related to *CCSS* in *ELA*.

Who Is in Charge?

Policy questions related to implementation often create anxiety. In addition to questions about assessment, questions about implementation timelines, course mapping, and teacher certification usually arise as teachers and administrators are introduced to *NGSS*. As a professional development facilitator, you must understand the decision-making process within your state. Is it centralized or do districts have a lot of local control? For example, in Wisconsin, the Department of Public Instruction makes the decision as to which standards to adopt. Individual districts have control over curriculum, including course mapping and sequencing, and curricular resources. However, course mapping may cause issues with teacher certification—an issue that is the responsibility of the state legislature.

Questions about implementation timelines also need to be addressed. We have found that many teachers have a perception that the timeline is "immediate"—that they will be expected to have *NGSS* fully implemented within a few months. This is unreasonable and dangerous; a hasty implementation will result in maintaining the status quo (at best). Implementation should be done deliberately with a multiyear timeline. If possible, you should make sure that participants are clear about the implementation timelines in their district.

Mapping the *NGSS* into courses, especially in middle and high school, can also become a sticking point. Unless mandated at the state level, course sequences are usually decided at the district level. However, in some cases, schools within a district are able to offer unique courses or course sequences. As a professional development facilitator, it is helpful for you to know this decision-making process. The activities in this book are useful for beginning the discussion of what courses look like but do not provide a full curricular design process.

What About Resources?

Some teachers are concerned with identifying resources, textbook series, and other curricular resources that are aligned with *NGSS*. In fact, we received one comment on an evaluation form that read, "When will the box of stuff come that shows us how to do this?" Publishers have been working at aligning and revising their materials for the *NGSS*, but no single resource will be perfect. Over the next few years, we will see a steady flow of new "*NGSS*-aligned" curricular resources. It will be critically important that teachers have a strong understanding of the vision and technical aspects of the standards as they choose resources. A strong implementation timeline and plan requires that teachers be mindful of the necessity to identify and review resources from a variety of sources.

Other teachers are concerned that they will need to "throw out" the materials they have been using to teach science. In this case, it is important for them to understand that *NGSS* represents an evolution of our understanding of standards, not a complete break with the past. Many of the resources based on inquiry or engineering design approaches will still be valuable for *NGSS* implementation. A tension often exists with budgets and textbook adoption cycles. In these cases, it is important to support teachers as they capitalize on existing resources. Activities in this book help teachers start the discussion about how to develop *NGSS* units and identify ways to emphasize practices and crosscutting concepts within existing materials. The Educators Evaluating the Quality of Instructional Products (EQuIP) Rubric, provided at *http://nstahosted.org/pdfs/ngss/EQuIPRubric.April.2014.pdf*, can provide further guidance.

We Already Do This!

This statement often crops up on initial examination of the science and engineering practices. A reading of the titles of the practices or a quick scan of the foundation boxes can lead to a superficial understanding of the *NGSS*. For example, everyone has students ask questions. Teachers' capacity for using the practices will vary in sophistication, but no teacher will be a blank slate. As a professional development facilitator, it is important that you acknowledge the good things that participants are already doing, foster a sharing of expertise, provide opportunities for participants to dig deeper into the details of each science and engineering practice, and provide options for participants to develop and practice new strategies.

Is Less More?

Questions related to rigor and amount of content can be challenging to address. Often, teachers become frustrated that content that they teach is not found in the *NGSS*. In some cases, it may be possible for teachers to include that content as examples to illustrate concepts in the standards. The *NGSS* writers tackled the "mile-wide and inch-deep" problem of U.S. science instruction head on. It is important to think carefully about what content we add to make sure that we do not squeeze out the space needed to have students engage content meaningfully and with depth. The *NGSS* Appendix A, "Conceptual Shifts in the *Next Generation Science Standards*" (and the activities in this book) provides evidence about the importance of depth over breadth. We need to redefine rigor as the quality of thinking, not the quantity of coverage.

Teachers of advanced or elective courses in high school also often become concerned when they do not see advanced concepts in *NGSS*. In these cases, professional development facilitators need to be clear that *NGSS* describes what *all* students should understand and be able to do at the end of their K–12 career. It is a baseline that all students should reach, but it should not constrain concepts in advanced courses. According to NSTA's Ted Willard, "*NGSS* sets the stage for Advanced Placement courses, but it does not replace them." In fact, many high school teachers have noted that the shifts apparent in *NGSS* are similar to those made in revisions to Advanced Placement guidelines.

4

TIPS FOR ADMINISTRATORS

As educational leaders, we constantly look for the magic formula—those "abracadabra" words that will magically make meaningful change happen with the snap of a finger or the wave of a wand. As we have all found out, those words do not exist. Change happens in a very thoughtful, planned way. Implementing and sustaining the *Next Generation Science Standards* (*NGSS*; NGSS Lead States 2013) in your district or school will not happen by magic; it will happen through thoughtful, careful, and planned hard work. As an administrator, you have the potential to greatly influence this process.

Support the Implementation of *NGSS* in Your District or School

Articulate your vision for science education and professional development to support science education. Administrators should be able to explain in-depth why *NGSS* are important to the school or district and to regularly include mention of this in public appearances (such as staff meetings) and informal discussions (such as classroom teacher conversations). Professional development can succeed only in settings or contexts that support it. Probably the most critical part of that support must come from administrators (McLaughlin and Marsh 1978). The outcome of your teachers' work on implementation of *NGSS* will ultimately depend on whether the administrators consider it important.

Professional Development Is Not an Event, It Is a Process

Professional development opportunities need to be made a regular part of the school year, happening over the summer or early in the year with periodic check-ins throughout winter and spring. "The most direct route to improving mathematics and science achievement for all students is better mathematics and science teaching," (NCMST 2000, p. 7). Helping teachers make this leap from learning something new to implementing it in the classroom is one of the most difficult pieces of professional development. Tom Guskey (2000) stated, "If there is one thing on which both behaviorists and cognitivists agree, it is that no one expects new learning to transfer immediately into more effective practice" (p. 180). Administrators need to provide support for the process so that transfer can take place.

Provide Ample Time for Teachers to Grow and Learn

As we all know, learning is a lifelong process. Teachers need time to learn, process, and put professional development into action. Giving teachers time to meet as a team to design a science curriculum is paramount to their understanding the process. Learning something new can be stressful. Teachers will need time to digest, reflect, and work with new content. It's during these periods when teachers can take time to process more deeply that they often achieve their "aha!" moments. Martin-Kniep (1999) believes that in order for teachers to sort out the information, make sense of it, and plan appropriate next steps, they must take time to reflect.

Time is a valuable resource, so make sure to use it wisely. A new program implementation can succeed only through a long-term approach. The fact is that a commitment to changing the way science is taught in your school is a commitment to continual improvement. It is not something you do and are done with. We can always improve. Quality professional development in science, or any other discipline, is never ending and ever evolving.

Be Visible and Participate in the Process

Administrators need to participate in the process of the implementation of *NGSS*. This not only includes participating in the professional development but all of the steps along the way: determining what you want to see and articulating your expectations to everyone, forming the right team to move the process forward, learning about

NGSS alongside your teachers, and giving teachers feedback on classroom implementation and if the *NGSS* are being taught with fidelity.

Build a Library of Resources

Provide teachers with the resources necessary for full implementation of *NGSS* in their classrooms. We especially recommend visiting the NGSS@NSTA Hub at *http://ngss.nsta.org*.
Resources could include

- web sites for additional information;

- video collections: quality classroom instruction, previous professional development, curriculum design process;

- resource books; and

- sample lessons, support materials, tips, and practical ideas.

Make Sure That All Staff Members, Not Just Your Science Teachers, Understand the Connection(s) Between NGSS and Classrooms

Teachers need to see the relevance of this work if they are going to internalize it and make it part of their being. Administrators must make explicit the connection between the expectations in NGSS, the needs of the schools or district, and the aspirations of teachers (Senge et al. 2000).

Maintain a Continuity of Focus Rather Than Having a Series of Divergent Workshops

As Ed Joyner stated in *Schools That Learn*, professional learning and professional work means no more "drive-by staff development." Staff development should not be disconnected from the core work of schooling. In this case, we are advocating for a focus on science education and the implementation of *NGSS*; not for having a series of disconnected staff development opportunities but a series of thoughtful focused opportunities to learn together. Teachers and administrators need to work together to solve problems and issues the district is having with science instruction and curriculum.

Understand and Embrace Collaboration

Teachers are used to working independently in their individual classrooms. However, studies have shown that there is a strong relationship between teacher collaboration and optimal professional growth. Collaboration allows teachers to gain multiple perspectives, encourages

self-reflection and good practices, and allows teachers to gain information that is relevant to them (Drago-Severson 2007). Plan activities that allow teachers to collaborate and work on implementing *NGSS* as a team.

What You Do as a Leader Matters!

Successful schools have successful leaders. Leithwood and his colleagues found that school leadership was second only to classroom instruction in the amount of impact on student learning (Leithwood et al. 2004). You, as the leader, need to lead learning. "To lead learning means to model a 'learner-centered,' as opposed to an 'authority-centered' approach to all problems, inside and outside the classroom" (Senge et al. 2000, pp. 416–417). In a learner-centered environment, all people in the system are viewed as learners and act as learners.

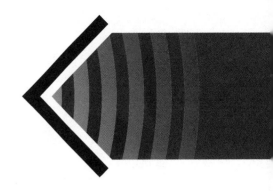

5

INTRODUCING THE *NGSS*

"Facts are not science—as the dictionary is not literature."
—Martin H. Fischer (1944)

Knowledge of science and engineering is important for all. An opening statement in *A Framework for K–12 Science Education* (*Framework*; NRC 2012) explains,

By the end of the 12th grade, students should have gained sufficient knowledge of the practices, crosscutting concepts, and core ideas of science and engineering to engage in public discussions on science-related issues, to be critical consumers of scientific information related to their everyday lives, and to continue to learn about science throughout their lives. They should come to appreciate that science and the current scientific understanding of the world are the result of many hundreds of years of creative human endeavor. It is especially important to note that the above goals are for all students, not just those who pursue careers in science, engineering, or technology or those who continue on to higher education. (p. 9)

In the 15 years since the National Research Council and the American Association for the Advancement of Science released the *National Science Education Standards*, there have been many changes in the world of science. In addition, there has been extensive research released on how students learn science. These pieces have been a driving force behind the writing of the *Framework* and the *Next Generation Science Standards* (*NGSS*; NGSS Lead States 2013).

The activities in this chapter provide an introduction to the *NGSS*. Activities 2, 3, and 4 represent three different ways to introduce the terminology and structure of the standards to educators. We do not intend that you do all three of these activities with the same group of teachers. Instead, choose the activity that best fits your presentation style.

ACTIVITY 1

Educators explore the progressions of *NGSS* to begin understanding the structure of *NGSS*.

ACTIVITY 2

Educators participate in a lecture and then discuss the development and structure of *NGSS* to begin developing a working knowledge of *NGSS*.

ACTIVITY 3

Educators use inquiry to develop their working definitions of *NGSS* vocabulary and the structure of *NGSS*.

ACTIVITY 4

Educators examine six conceptual shifts in *NGSS* that demonstrate how *NGSS* is different from previous standards documents.

ACTIVITY 1

Examining the Standards

Approximate Length

30–40 minutes

Objectives

During this activity, participants will

- explore one strand of the standards from early elementary through high school,

- describe how the standards' expectations progress as the grade level increases, and

- summarize their discussions on chart paper.

Vocabulary

- progressions

- *NGSS* standards page

- performance expectations

- foundation boxes

Evidence of Learning

- Group summary on chart paper

- Graphic organizer "Examining the Standards"

At a Glance

In this activity, participants explore one strand (organized by either topic or disciplinary core idea) of the standards from early elementary through high school. This activity works well for helping educators get a general feel for the *NGSS* and develop an understanding that the content in the standards builds developmentally over the course of a student's education.

When we design professional development, we try to use a learning cycle approach whenever possible. This often means that we use an "ABC" or "activity before content" format to give participants the chance to engage and explore concepts before we explicitly introduce the content of the session. When time permits, we follow the introduction of content with an opportunity for participants to apply that content. We have found that this activity works well for introducing *NGSS* before we dig in to the development and structure of the standards. Participants only need a limited understanding of how to read an *NGSS* standards page. This is not a stand-alone activity. It should be followed by an activity that introduces the purpose and structure of *NGSS* (e.g., Activity 2, 3, or 4 in this book).

Facilitator's Notes

Since this is an exploration activity, do not focus on providing a comprehensive overview of the *NGSS* or how to read a standards page. Participants quickly notice that the content at each grade level builds on the previous level and introduces increased complexity without being redundant. Most participants are excited by the clarity of the verbs used in the performance expectations (the science and engineering practices) and note that students are expected to be able to "do things with the content they are learning."

Materials

- Copies of the handout "Examining the Standards" (p. 27)

- Chart paper and markers (per group of three to four)

- Standards progression (per group of three to four). Identify one set of related standards that includes standards in grades K–2, 3–5, 6–8, and 9–12. For example, we have used a "waves progression" (organization by topic) that included the following pages: 1. Waves: Light and Sound; 4. Waves: Waves and Information; MS. Waves and Electromagnetic Radiation; HS. Waves and Electromagnetic Radiation.

Procedure

Set-up: Participants should be organized into small groups prior to starting this activity. If possible, place participants in mixed grade-level groups.

Introduction (5 minutes): After giving participants a copy of the handout and standards progression, provide a very brief introduction to reading an *NGSS* standards page. Explain that both the standards page title and performance expectation code (e.g., 1-PS4-1) identify the grade level. Also state that the foundation boxes in the middle of the page provide more depth as to what students should know and be able to do at each grade level. You do not need to provide a comprehensive overview of the standards page at this time.

Group Work (20 minutes): Charge the participants to explore the standards that you have given them. They should pay particular attention to how expectations progress as the grade level increases. Participants should complete the organizer on the handout. With five minutes remaining in this stage, instruct participants to summarize their discussions on a chart paper. The summary should include questions that they have about the standards. At the end of this stage, each group should post their summary so that it is visible for the entire group.

Debrief (5–10 minutes): Ask a few of the groups to present their summary to the whole group. Foster cross talk between groups by asking participants to describe differences between their summary and previous presentations.

Wrap-up (5 minutes): Conduct a gallery walk by giving participants a chance to look at the group posters. Participants should place a check next to questions on the summaries that resonate with them.

Next Steps

If this is your participants' initial exposure to *NGSS*, consider following this activity with Activity 2, 3, or 4 to introduce the background and structure of the standards. You should also make note of the questions on the group summary posters. Many of these questions can be answered by using the activities in this book. However, to answer questions related to state or district policy, you will need additional resources.

ACTIVITY 1

Examining the Standards

Examine a *progression* of standards from *NGSS*. As you examine it, think about the questions below. It may be useful to compare the *NGSS* to your existing standards.

How does the content build over grade levels?	How are higher-order thinking skills integrated into these standards?	
Is it clear what students are expected to know and be able to do?	**What excites you about these standards?**	**What concerns or "wonderings" do you have about these standards?**

ACTIVITY 2

NGSS Vocabulary

Approximate Length

45 minutes

Objectives

During this activity, participants will

- learn the structure of *NGSS*,

- read a standards page, and

- explore the standards for their grade level or grade band.

Vocabulary

- *Next Generation Science Standards* (*NGSS*)

- *Common Core State Standards* (*CCSS*)

- *A Framework for K–12 Science Education*

- science and engineering practices

- crosscutting concepts

- disciplinary core ideas

- performance expectations

- assessment boundaries

- clarification statements

- foundation boxes

- connection boxes

Evidence of Learning

- Lists of generated questions

At a Glance

This activity is a straightforward presentation about the development and structure of the *NGSS*. The presentation ends by providing an overview of a standards page. "*NGSS* Vocabulary" is one of three introductory presentations included in this book.

Facilitator's Notes

The following narrative provides the background information needed for this activity. Facilitators should feel free to determine the best way to present this information. For example, you might present it as a PowerPoint presentation, a lecture, or a jigsaw reading.

Then and Now

Much has happened since the *National Science Education Standards* (*NSES*) were released by the National Research Council in 1996. Putting the previous standards in context can be helpful for understanding why new standards are needed.

Science and Technology

When the *NSES* were released, "soccer mom" was the word of the year, and "dot" (as in dot-com) was selected as the most useful word of the year. We had to worry about rewinding VHS tapes before returning them to the rental store and had trouble jogging without our compact disc players skipping. Hitachi released a camcorder that could take both still and moving digital pictures. For $2,000, your camera could store 20 minutes of video or 3,000 pictures, with a stunning 0.3 megapixel resolution. Shortly after the *NSES* were released, the first personal digital music player hit the market: The $400 MPMAN by SaeHan could store six of your favorite songs. Dolly the sheep had not yet been cloned (that happened July 5th, 1996). We only had 111 elements (now we have 118). Mars Pathfinder was waiting for launch (it began its journey December 4, 1996). And the human genome had not been mapped.

Science Education

We have also learned a lot about teaching and learning in science. The journals *Science Education*, the *Journal of Research in Science Teaching*, and the *International Journal of Science Education* alone have published more than 2,500 peer-reviewed research articles. Achieve Inc. has published the *International Science Standards Benchmarking Report* and the National Research Council (NRC) has published multiple reports on effective science education, including the following:

- *How Students Learn History, Science, and Mathematics in the Classroom*

- *America's Lab Report*

- *Taking Science to School*

- *Ready, Set, Science!*

- *Successful STEM Schools*

Perhaps it is time to update our standards.

Development

This activity is not intended to go in depth about the development process for the *NGSS*. However, it may be important to provide a few highlights of the process.

The *NGSS* are not part of the *CCSS* initiative. *CCSS* should only be used to refer to the *CCSS*, in English language arts (ELA) and mathematics.

The *CCSS ELA* does include an appendix of literacy standards for science and technical fields. These standards guide disciplinary literacy standards for science teachers in grades 6–12 but do not include science content standards.

The development of *NGSS* included a partnership between the National Science Teachers Association, the National Research Center, the American Association for the Advancement of Science, and Achieve Inc. *NGSS* is not an initiative of the U.S. Department of Education or any other federal agency. The Carnegie Foundation funded the development.

The first step of the development process culminated in the publication of the *Framework* by NRC. The *Framework* was the guiding document for the *NGSS* writing team, which was managed by Achieve Inc. This writing team was composed of classroom teachers, scientists, and science education researchers. The development process included a comprehensive review process with teams from multiple states and several public review periods.

For more information on the review process, visit the *NGSS* website at *www.nextgenscience.org/development-overview*.

Framework

The NRC's *Framework* provides a vision for what science should look like in the United States. The *Framework* defined the following three dimensions: science and engineering practices, crosscutting concepts, and disciplinary core ideas.

Science and Engineering Practices

The *Framework* identified eight science and engineering practices. These are intentionally called practices (instead of skills) to acknowledge that in order to engage in the process of science and engineering, students need to have specific knowledge. The eight science and engineering practices build developmentally from kindergarten all the way through high school. The practices are as follows: asking questions and defining problems; developing and using models; planning and carrying out investigations; analyzing and interpreting data; using mathematics and computational thinking; constructing explanations and designing solutions; engaging in arguments from evidence; and obtaining, evaluating, and communicating information. It is important to note that the final three practices are very well aligned and complementary to the *CCSS* for English Language Arts and Literacy in History/Social Studies, Science, and Technical Subjects.

Crosscutting Concepts

The *Framework* identifies seven crosscutting concepts. Crosscutting concepts are the concepts or ideas that stretch across all disciplines of science. The benefit of focusing on crosscutting concepts is that it helps provide students with an organizational structure for understanding the world. The crosscutting concepts build from kindergarten through twelfth grade and include patterns; cause and effect; scale, proportions, and quantity; systems and system models; energy and matter in systems; structure and function; and finally, stability and change in systems.

Disciplinary Core Ideas and Component Ideas

The *Framework* outlines a series of 13 disciplinary core ideas. These core ideas are foundational to science. There are four core ideas each in physical science and life science. There are three core ideas in the Earth and space sciences and two

engineering core ideas. Component ideas provide additional detail for each core idea.

One of the goals of the *Framework* was to identify a coherent scope and sequence of a few ideas that are central to science and build on them throughout a student's K–12 education career.

Performance Expectations

The standards are written as learning progressions that integrate disciplinary core ideas, science and engineering practices, and crosscutting concepts. Performance expectations serve as guidelines for assessment, not instructional tasks or curriculum mandates. Many performance expectations also include assessment boundaries and clarification statements to further define appropriate depth at that grade level or grade band.

Reading a Standards Page

An initial look at a standards page from *NGSS* can be quite overwhelming. However, as educators gain comfort with *NGSS*, they can begin to see how the different pieces of a page work together. A cluster of performance expectations related to a specific topic or core idea sits at the top of the page. The three foundations boxes are found directly beneath the performance expectations. These three foundations boxes contain statements, many directly from the *Framework*, that further define student learning expectations for each of the three dimensions. A series of connections boxes can be found at the bottom of the page. Connections boxes illustrate how the performance expectations on that page are related to other performance expectations within *NGSS* and provide connections to the *CCSS*. These connections are included as a starting point to determine how mathematics and literacy concepts and skills can be integrated or reinforced during science instruction.

Materials

- Each participant should have access to at least one standards page. You should decide in advance if you will be using the *NGSS* organized by topics or by disciplinary core ideas.

Procedure

Introduction (5–10 minutes): Begin this activity by activating prior knowledge. Ask participants to reflect on the question, "What is the purpose of curriculum standards?" Provide an opportunity for participants to briefly discuss this question in small groups. End the introduction by doing a round robin, in which each group suggests one idea related to the question.

Presentation (20 minutes): Use the Facilitator's Notes to provide participants with an overview of the development and structure of the *NGSS*.

Explore (10 minutes): During this step, participants should be given time to explore the standards for their own grade level or band. As they explore the standards, encourage them to generate and record questions.

Debrief (10 minutes): Close this activity with a question and answer session. You may be able to address some of these questions directly; other questions may be answered by using activities in this book. Finally, some questions that are asked may be specific to state or district policy decisions. If you do not know how to answer these questions, make sure that you do not speculate.

ACTIVITY 3

The Structure of *NGSS*

Approximate Length

55 minutes

Objectives

During this activity, participants will

- define the different structural parts of *NGSS*,

- learn the structure of *NGSS*,

- read a standards page, and

- explore the standards for their grade level or grade band.

Vocabulary

- *Next Generation Science Standards (NGSS)*

- *Common Core State Standards (CCSS)*

- *A Framework for K–12 Science Education*

- science and engineering practices

- crosscutting concepts

- disciplinary core ideas

- performance expectations

- assessment boundaries

- clarification statements

- foundation boxes

- connections boxes

Evidence of Learning

- Graphic organizer describing the components of *NGSS*, "The Structure of *NGSS*"

- List of generated questions

At a Glance

This is one of three activities in this book that can be used to introduce educators to the structure of the *NGSS*. Instead of providing the structure through a lecture (as in Activity 2), participants use one of the introductory sections of *NGSS* to determine their own definitions for important elements of the *NGSS* structure (performance expectations, disciplinary core ideas, foundation boxes, connections boxes).

Facilitator's Notes

See the Facilitator's Notes on pages 28–31 in Activity 2 for the background information necessary to facilitate this activity.

Materials

- Each participant should have a copy of the "*NGSS* Structure" document from the front matter of the *NGSS*. (This document can be found at *www.nextgenscience.org/ next-generation-science-standards*.)

- Copies of the handout "The Structure of *NGSS*" (p. 35)

- Access to at least one *NGSS* standards page

Procedure

Before beginning this activity you should be comfortable with the background information on the development and structure of *NGSS* provided in the Facilitator's Notes of Activity 2 (pp. 28–31).

Set-up: If you are going to use small groups, create them prior to beginning this activity. This activity works best if you put participants into pairs or groups of three. Groups can be either mixed or grade-level teams.

Introduction (5–10 minutes): Begin this exercise by activating prior knowledge. Ask participants to reflect on the question, "What is the purpose of curriculum standards?" Provide an opportunity for participants to briefly discuss this question in small groups. End the introduction by doing a round robin during which each group suggests one idea related to the question. (*Note*: This introduction is the same as that in Activity 2, p. 31).

Group Work (20 minutes):
Place participants into pairs or small groups and distribute materials. Explain to the participants that they will be using the "*NGSS* Structure" section of the *NGSS* to define concepts that are important to understanding the standards. *Optional*: You may want to begin this activity by providing context from the Facilitator's Notes in Activity 2 (pp. 28–31). Participants should use the next 15 minutes to read the "*NGSS* Structure" document and come to a consensus on how to answer the questions in the handout.

FIGURE 5.1

Model presentation figure

THE STANDARD
Title:
Performance expectations describe what students should know and be able to do at the end of instruction. Performance expectations guide summative and formative assessment.

The foundation boxes provide the context for performance expectations.		
Science and engineering practices	**Diciplinary core ideas**	**Crosscutting concepts**

The connection boxes provide guidance for connecting the standard to others in *NGSS* or the *CCSS*.

Note: *Disciplinary Core Ideas form the main concepts that are essential to the major science disciplines. These 39 ideas are drawn from* A Framework for K–12 Science Education *and span kindergarten through grade 12.*

Report Out (10 minutes): Ask groups to report out how they have answered the questions on the handout. Stimulate cross talk between groups by encouraging them to share how their answers are similar and different from each other. Clarify definitions for performance expectations, disciplinary core ideas, the components of the foundations boxes, and the connections boxes as needed. Display Figure 5.1 as a summary of the *NGSS* structure.

Explore (10 minutes): During this step, participants should be given time to explore the standards for their grade level or band. As they explore the standards, encourage participants to generate and record questions.

Debrief (10 minutes): Close this activity with a question and answer session. You may be able to address some of these questions directly; other questions may be answered by using activities in this book. Finally, some questions that are asked may be specific to state or district policy decisions. If you do not know how to answer these questions, make sure you do not speculate.

ACTIVITY 3
The Structure of NGSS

How would you describe an *NGSS* performance expectation to a colleague?

What is a disciplinary core idea?

What is the purpose of the foundation boxes?

What are the connections boxes?

ACTIVITY 4

NGSS Conceptual Shifts

With Chad Janowski

Approximate Length

55 minutes

Objectives

During this activity, participants will

- come to understand the rationale for the conceptual shifts in *NGSS*,

- reflect on how these shifts benefit student learning, and

- reflect on how these shifts positively impact their instructional planning.

Vocabulary

- conceptual shift

- three dimensions

- progressions

- *Next Generation Science Standards* (*NGSS*)

- *Common Core State Standards* (*CCSS*)

- *A Framework for K–12 Science Education*

- science and engineering practices

- crosscutting concepts

- disciplinary core ideas

- performance expectations

- assessment boundaries

- clarification statements

- foundation boxes

- connections boxes

Evidence of Learning

- Responses to handout "Conceptual Shifts in the *NGSS*"

At a Glance

This is one of three activities that can be used to introduce the development and structure of the *NGSS* to educators. This activity focuses on the six conceptual shifts that demonstrate how *NGSS* is different from previous standards documents. This activity serves three purposes:

- To provide a rationale for the importance of the conceptual shifts

- To give teachers a chance to discuss the impact of the conceptual shifts

- To illustrate how these shifts are reflected on a standards page as a way to help participants learn how to read the *NGSS*

Facilitator's Notes

Educators and education researchers have learned a lot about designing effective standards since the release of the *National Science Education Standards* more than 15 years ago. As a result, the *NGSS* writing team made a series of six conceptual shifts. Understanding those shifts, and how they are reflected in the structure of the standards, is important to understanding the vision of the *Framework* and the standards.

The first and fifth conceptual shifts reflect how science and engineering are done in the real world by integrating content, practices, and crosscutting concepts while raising the profile of engineering in science education. The National Research Council's *America's Lab Report* (2005) is a synthesis of research related to the efficacy of science laboratory activities. One major finding was that integrated learning experiences increase student understanding and transfer of that understanding to different situations. We also know that context and the integration of content, practices, and crosscutting concepts supports the learning of students from nondominant groups and helps English language learners develop language skills (see Chapter 7).

From work in neuroscience, we also know that

experiential learning that stimulates multiple senses in students, such as hands-on science activities, is not only the most engaging but also the most likely to be stored as long-term memories. … The best-remembered information is learned through multiple and varied exposures followed by authentic use of the knowledge. (Willis 2006, p. 6)

These two shifts will positively impact how students experience and learn science in our classrooms.

The third and fourth shifts focus on the need for coherency and a focus on depth of understanding of core ideas. One common criticism of U.S. science education is that we try to cover large amounts of content without providing time to develop an understanding of concepts; we give our students fat textbooks and race to cover them during the school year. Phil Sadler and colleagues (Tai, Sadler, and Mintzes 2006) found that high school students who study a topic in depth for one month are much more successful during introductory university science courses when compared with students in courses characterized by covering many more topics.

Wiggins and McTighe (2011) also point to research that shows that deeper understanding is important. They say

research on expertise suggests that superficial coverage of many topics in the domain may be a poor way to help students develop the competencies that will prepare them for future learning and work. Curricula that emphasize breadth of knowledge may prevent effective organization of knowledge because not enough time is provided to learn anything in depth. Curricula that are 'a mile wide and an inch deep' risk developing disconnected rather than connected knowledge. (p. 5)

The first and sixth shifts are related to the use of performance expectations and the inclusion of connections to the *CCSS*. Performance expectations highlight the importance of integrating the dimensions and call for an emphasis on performance assessments. Wiggins and McTighe (2011) state,

Many assessments measure only recently taught knowledge and never ask for authentic performance (conditional knowledge and skills)—whether students know when, where, and why to use what they have learned in the past. This approach leads to surprisingly poor test results, because students do not recognize prior learning in unfamiliar-looking test questions—especially when the test has no context clues and hints (as occurs when teachers immediately quiz students on recent material). (p. 5)

Additionally, strong connections to the *CCSS* will help teachers of science better align instruction with what we know about disciplinary literacy and how to reinforce mathematical concepts and skills.

Materials

- Copies of the handout "Conceptual Shifts in *NGSS*" (pp. 39–42)

- Each participant will also need access to the *NGSS*

Procedure

Before beginning this activity, review the background information provided in the Activity 2 Facilitator's Notes (pp. 28–31) and read *NGSS* Appendix A, "Conceptual Shifts in the *Next Generation Science Standards.*"

Introduction (5–10 minutes): Begin this activity by activating prior knowledge. Ask participants to reflect on the question, "What is the purpose of curriculum standards?" Provide an opportunity for participants to briefly discuss this question in small groups. End the introduction by doing a round robin in which each group suggests one idea related to the question. (*Note*: This introduction is the same as in Activity 2.)

Presentation (35 minutes): You may want to begin by using the Activity 2 Facilitator's Notes to provide context for why the *NGSS* are important.

Provide each participant with the handout "Conceptual Shifts in the *NGSS*."

Use the Facilitator's Notes from this activity to provide participants with background on the conceptual shifts and structure of *NGSS*. In this activity, we clustered the conceptual shifts into three pairs. After you discuss each pair, pause to give participants time to answer the associated question. You may want to use a Think-Pair-Share strategy at this point. After the question has been answered, you can continue the presentation by showing participants how the conceptual shift is reflected in the *NGSS* structure and on an *NGSS* standards page.

Explore (10 minutes): During this step, participants should be given time to explore the standards for their grade level or band. As they explore the standards, encourage participants to generate and record questions.

Debrief (10 minutes): Close this activity with a question and answer session. You may be able to address some of these questions directly; other questions may be answered by using activities in this book. Finally, some questions that are asked may be specific to state or district policy decisions. If you do not know how to answer these questions, make sure you do not speculate.

ACTIVITY 4

Conceptual Shifts
in the *NGSS*

See NGSS *Appendix A for more details on the conceptual shifts. The shift descriptions in this activity are from the article "NGSS Conceptual Shifts" by Karen L. Ostlund.* NSTA Reports, *February 28, 2013.* www.nsta.org/publications/news/story.aspx?id=59853.

Pair 1

Shift 1: The *NGSS* reflect how science is done in the real world by intertwining three dimensions: scientific and engineering practices, crosscutting concepts, and disciplinary core ideas. Scientists ask and answer questions to further our understanding of the world around us. Engineers define problems and design solutions to solve problems. The intent of *NGSS* is to weave the three dimensions together to reflect the work of scientists and engineers. For example, students are expected to use scientific and engineering practices and apply crosscutting concepts to develop an understanding of disciplinary core ideas. This is a conceptual shift from most state and district standards, which separate these dimensions in curriculum, instruction, and assessment. Curriculum often initially focuses on the science process skills of inquiry without emphasizing science content. To prepare students for the competitive global economy, we must equip them with the skills and information to develop a sense of contextual understanding of scientific knowledge: how scientists acquire it, how engineers apply it, and how it is connected through crosscutting concepts. These understandings can be achieved by interlocking the three dimensions. Therefore, each *NGSS* performance expectation integrates scientific and engineering practices to understand disciplinary core ideas and connect ideas across disciplines by applying crosscutting concepts.

Shift 5: The *NGSS* integrate science, technology, and engineering throughout grades K–12. The *NGSS* integrate applications of science, technology, and engineering into the disciplinary core ideas along with life, Earth, space, and physical science. This conceptual shift also raises engineering design to the same level as scientific inquiry. This requires the development of curriculum, instruction, and assessments—as well as teacher preparation—to integrate engineering and technology into the structure of science education. Science and engineering are needed to address challenges we face in our ever-changing world, such as an adequate food supply, clean water, renewable energy, and disease control. Hopefully, students will be motivated to pursue careers rooted in science, technology, engineering, and mathematics as a result of early opportunities to apply their scientific knowledge to develop solutions to similar challenges. Integrating science, technology, and engineering into curriculum and instruction empowers students to apply what they learn to their everyday lives beginning in kindergarten, throughout their academic careers, and beyond.

How will these shifts benefit student learning in your classroom?

Pair 2

Shift 3: The *NGSS* build coherently from grades K through 12. The *NGSS* concentrate on a limited number of essential disciplinary core ideas that build student understanding progressively from grades K through 12. The conceptual shift is the movement away from learning disjointed and isolated facts and toward opportunities to learn more complex ideas as students progress through grade levels and bands. The disciplinary core ideas identified in *NGSS* form a coherent progression of knowledge leading to more complexity of student understanding by the end of high school. The goal is to help students achieve scientific literacy by focusing on fundamental content that builds as they progress. The *NGSS* progressions are based on the assumption that students have learned previous content and can build on their understandings. Therefore, it is critical that students master the content designated for each grade level or band. The omission of any content in a grade level or band can negatively impact student understanding of increasingly more complex core ideas as students advance through grades K–12.

Shift 4: The *NGSS* focus on deeper understanding of content and applications of content. The *NGSS* focus on disciplinary core ideas rather than the myriad of facts associated with each core idea. Although the facts support the core ideas, they should not be the focus of curriculum, instruction, and assessment. The conceptual shift places more emphasis on the core ideas and less on facts to provide an organizational structure that delivers the scaffolding students need when acquiring new knowledge. Research indicates experts understand core principles and theoretical constructs of their field and use them to make sense of new information or to apply their understandings to solve problems. Novices hold disconnected and even contradictory pieces of knowledge as isolated facts and have difficulty organizing and integrating the pieces. Therefore, the intent of the *NGSS* is to engage students in scientific and engineering practices to gain a deeper understanding of disciplinary core ideas and connect those ideas with crosscutting concepts to help them develop from novices into experts.

How will these shifts benefit student learning in your district?

Pair 3

Shift 2: The *NGSS* are student performance expectations. Student performance expectations clarify what students should know and be able to do at the end of a grade level or band. This conceptual shift recognizes the *NGSS* are *not* curriculum, instruction, or assessment; the *NGSS* *are* student performance expectations that elucidate the intent of assessments. The *NGSS* will guide curriculum developers as they develop coherent instructional programs designed to ensure students attain the performance expectations. The three dimensions are integrated in each performance expectation and are intended to enhance instruction and curriculum, not limit it. The scientific and engineering practices and crosscutting concepts should be used throughout the curriculum and instruction so students have many opportunities to become proficient at using the practices to deepen their understanding of disciplinary core concepts by connecting them with crosscutting concepts. Backward curriculum design will be required to analyze each performance expectation and develop an instructional sequence that will help students achieve the outcomes delineated in the *NGSS*.

Shift 6: The *NGSS* correlate to the *CCSS* in English language arts (ELA) and mathematics. The *NGSS* are the vehicle for mastering the *CCSS* in ELA and mathematics. Science and engineering provide a content area for applying ELA and mathematics skills. The conceptual shift is away from viewing ELA and mathematics as content areas to the perception that they are skills to be practiced and mastered in the science and engineering curriculum. A synergy is created when ELA, mathematics, science, and engineering standards reinforce the acquisition of the skills and knowledge in all of these areas of the school curriculum.

How will these shifts positively impact your instructional planning?

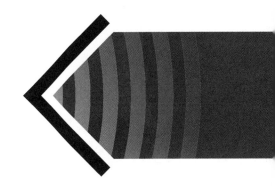

6

Exploring the Practices and Crosscutting Concepts

It is astonishing to realize that until Galileo performed his experiments on the acceleration of gravity in the early seventeenth century, nobody questioned Aristotle's falling balls. Nobody said, "Show Me!"
—Neil deGrasse Tyson (1994, p. 17)

A *Framework for K–12 Science Education* (*Framework*; NRC 2012) and the *Next Generation Science Standards* (*NGSS*; NGSS Lead States 2013) are built on three dimensions of science. Taken together, these dimensions—science and engineering practices, disciplinary core ideas, and crosscutting concepts—provide students with a deep understanding of science as a way of thinking about the world and as an accumulated body of knowledge about our world. The *Framework* is very clear on the purpose of integrating these dimensions toward developing scientifically literate citizens:

By the end of the 12th grade, students should have gained sufficient knowledge of the practices, crosscutting concepts, and core ideas of science and engineering to engage in public discussions on science-related issues, to be critical consumers of scientific information related to their everyday lives, and to continue to learn about science throughout their lives. They should come to appreciate that science and the current scientific understanding of the world are the result of many hundreds of years of creative human endeavor. It is especially important to note that the above goals are for all students, not just those who pursue careers in science, engineering, or technology or those who continue on to higher education. (p. 9)

The activities in this chapter will help educators begin to understand these three dimensions and how they can be integrated into rich learning experiences for students.

ACTIVITY 5

Educators experience a model activity and identify how practices, core ideas, and crosscutting concepts are integrated into instruction.

ACTIVITY 6

Educators explore the *Framework*'s definition of science inquiry as scientists (and students) engaging in the practices across three spheres of activities. Educators also examine how the practices provide clear and assessable statements as to how students should be able to engage with scientific content and processes. This is an extension of the previous activity.

ACTIVITY 7

Educators experience an engineering design activity and explore the practices and the *Framework*'s spheres of activities from an engineering perspective.

ACTIVITY 8

Educators complete a self-assessment related to the practices of science and engineering in order to determine strengths, weakness, and "next steps" for professional development.

ACTIVITY 9

Educators explore crosscutting concepts in the *NGSS* and reconstruct progressions to understand how they become more sophisticated over time. Educators also identify opportunities for building explicit connections between disciplinary core ideas and crosscutting concepts.

ACTIVITY 10

Educators participate in a model activity and reflect on the characteristics of the nature of science. They then construct draft anchor charts to help students understand these characteristics throughout the school year.

ACTIVITY 5

Integrating the Three Dimensions

Approximate Length

180–220 minutes

Objectives

During this activity, participants will

- identify disciplinary core ideas, science and engineering practices, and crosscutting concepts in a model activity;

- revise a model activity to focus on one or more practices and crosscutting concepts;

- reflect on connections of the three dimensions in the model activity;

- identify a unit they already teach and plan to modify the unit to incorporate the three dimensions; and

- develop an action plan to increase student focus on the use of specific practices and crosscutting concepts.

Vocabulary

- crosscutting concepts

- disciplinary core ideas

- science and engineering practices

- three dimensions

Evidence of Learning

- Summarization of work on chart paper

- Graphic organizer "Integrating Crosscutting Concepts"

- Graphic organizer "Integrating Science and Engineering Practices"

At a Glance

In this activity, participants will experience a model activity as they develop an understanding of the three dimensions of the *Framework* and *NGSS*. After experiencing the model activity, participants will identify related disciplinary core ideas, science and engineering practices, and crosscutting concepts. This activity ends with participants suggesting ways that they could make instruction more explicit for one or more practices and crosscutting concepts. This activity assumes that participants have a beginning understanding of the structure of the *NGSS* and that the *NGSS* performance expectations were developed to include the three dimensions from the *Framework*.

As written, this activity introduces participants to both the crosscutting concepts and the science and engineering practices. However, Activity 6 uses the same model activity and digs deeper into the connection between science inquiry and the science and engineering practices. We also provide suggestions for modifications that will reduce overlap between these two activities.

Facilitator's Notes

Perhaps the largest and most exciting difference between the *NGSS* and other current standards is the integration of content, practices, and crosscutting concepts. The *NGSS* performance expectations guide assessment toward considering content within the context of using practices and making connections with crosscutting concepts. For example, the *NGSS* say that

state standards have traditionally represented Practices and Core Ideas as two separate entities. Observations from science education researchers have indicated that these two dimensions are, at best, taught separately or the Practices are not taught at all. This is neither useful nor practical, especially given that in the real world science and engineering is always a combination of content and practice. It is important to note that the Scientific and Engineering Practices are not teaching strategies—they are indicators of achievement as well as important learning goals in their own right. As such, the Framework and NGSS ensure the Practices are not treated as afterthoughts. Coupling practice with content gives the learning context, whereas practices alone are activities and content alone is memorization. It is through integration that science begins to make sense and allows student to apply the material. This integration will also allow students from different states and districts to be compared in a meaningful way. (p. 2)

In addition, the *Framework* states that

crosscutting concepts have value because they provide students with connections and intellectual tools that are related across the differing areas of disciplinary content and can enrich their application of practices and their understanding of core ideas. (p. 233)

Materials

- Provide copies of the following handouts: "Integrating Crosscutting Concepts" (p. 48) and "Integrating the Science and Engineering Practices" (p. 49).

- Provide access to the "Science and Engineering Practices" matrix (*NGSS*, Appendix F) or copies of the "Science and Engineering Practices Self-Assessments" handout from Activity 8 (pp. 70–80), the "Crosscutting Concepts" matrix (*NGSS*, Appendix G) and the "DCI Progressions" matrix (*NGSS*, Appendix E).

- This activity uses one of two model activities. The model activity for use with elementary audiences is the Explore section of "Classroom Curling: Exploring Forces and Motion" from *Inquiring Scientists, Inquiring Readers* (found in Appendix 2, p. 203). The model activity for use with secondary audiences is "Now You 'Sea' Ice, Now You Don't: Penguin Communities Shift on the Antarctic Peninsula" from *Climate Change from Pole to Pole: Biological Investigations* (found in Appendix 3, p. 209).

- Supply chart paper with the debrief questions (see Debrief section of this activity).

Procedure

Set-up: Participants should be organized into groups appropriate for the model activity you will be using. The "transfer" component of this activity works best when participants are in grade-level or content-alike groups. Materials for the model activity should be prepped in advance.

Introduction (5 minutes): Begin the activity by explaining that participants will first experience a model activity and then use this experience to identify how the three dimensions of *NGSS* can be integrated into instruction.

Model Activity (60 minutes): Participants should experience the model activities (i.e., Appendix 2 or 3) as students.

Model Activity Wrap-up (10 minutes): Transition participants from engaging in the model activity as a student back to thinking as a teacher by using a Think-Pair-Share approach focused on the question, "What aspects of good science teaching were evident in this model activity?"

Integrating the Dimensions (60 minutes): This component of the activity is divided into the following four parts:

Disciplinary Core Ideas (10 minutes): Participants should use the disciplinary core idea progressions in *NGSS* Appendix E to identify the disciplinary core ideas used in this activity.

Crosscutting Concepts (20 minutes): Pass out the handout, "Integrating the Crosscutting Concepts." Participants should use the "Crosscutting Concepts" matrix (*NGSS*, Appendix G) to identify the crosscutting concepts that are related to this activity. In addition, participants should select one crosscutting concept and describe how they could make the connection between the disciplinary core idea and crosscutting concept more explicit for students.

Science and Engineering Practices (20 minutes): (*Note*: If you are planning to use Activity 6 as an extension to this activity, eliminate this phase.) Pass out the "Integrating the Science and Engineering Practices" handout. Participants should use the "Science and Engineering Practices" matrix in Appendix F of the *NGSS* to identify the practices that they engaged in during the model activity. Participants should also describe what they were doing as they engaged in that practice. Finally, participants should select one practice and describe how they could make the use of that practice more explicit for students.

Gallery Walk (10 minutes): Participants summarize their work on chart paper that can be displayed. Provide groups with time to view each other's work. Encourage participants to comment in writing on the posters created by other groups.

Transfer (25 minutes): Participants work either in grade-level or topic-alike groups to identify an activity in an upcoming unit that they can modify to incorporate the three dimensions of the *Framework*. Participants should be encouraged to develop an action plan to increase student focus on the use of specific practices and at least one crosscutting concept during that unit. (*Note*: If you are using Activity 6 as an extension to this activity, you can either use this transfer phase after completing that activity or you can narrow the scope by having participants reflect only on crosscutting concepts.)

Debrief (40 minutes): Use a "knowledge café" approach to provide closure to this activity. In a knowledge café, participants engage in multiple informal discussions and record notes on chart paper. To conduct this closure, you will need to create groups of four to five people. These groups should be different than those used in previous parts of this activity. Each group will discuss the following three prompts:

- What are your two to three biggest insights from this activity?

- What are your two to three biggest concerns after completing this activity?

- What questions do you have?

Write one prompt at the top of each sheet of chart paper, and spread the prompts around the room. (*Note*: If you have more than three groups,

create extra sets of prompts on chart paper). Each group should start at a different prompt. After discussing their initial prompt (and recording their big ideas) for seven to eight minutes, groups should rotate to the next prompt. Groups should quickly read the notes from the previous group, discuss the prompt, and add their ideas to the chart paper. Continue until every group has discussed each prompt. After the groups discuss the last prompt, they should take a few minutes to summarize the notes from all groups related to that prompt. End the activity by posting the summarized chart paper and allow participants to browse.

Next Steps

This activity illustrates a practical classroom example of how disciplinary core ideas, science and engineering practices, and crosscutting concepts should be integrated into instruction. The remaining activities in this chapter help science teachers gain a deeper understanding of the crosscutting concepts and science and engineering practices. Activity 6 is an extension to this activity that digs deeper into science inquiry and the science and engineering practices. In addition, Activity 11 introduces the case studies in *NGSS* Appendix D, "All Standards, All Students." These case studies include wonderful classroom vignettes that show integration of disciplinary core ideas, science and engineering practices, and the crosscutting concepts in contexts that support the learning of students from nondominant populations.

ACTIVITY 5

Integrating Crosscutting Concepts

Describe the crosscutting concepts involved in this activity.

What could you do to help students make connections between the activity content and at least one crosscutting concept?

ACTIVITY 5

Integrating the Science and Engineering Practices

Identify the science and engineering practices involved in this activity. In addition, describe what you were doing as you engaged in each practice.

Science and engineering practice	What did it "look" like?

What could you do to help students develop one science and engineering practice in this activity?

ACTIVITY 6

Science Inquiry and the Practices of Science and Engineering

Approximate Length

60–90 minutes

Objectives

During this activity, participants will

- explore the connections between the *Framework*'s spheres of activity for scientists and engineers and the science and engineering practices,

- describe how a model activity engaged participants in the three spheres,

- evaluate a model inquiry activity for classroom use, and

- identify science and engineering practices used in the model activity.

Vocabulary

- inquiry

- modeling

- explanations

- argumentation

- three spheres (investigating, explaining or solving, evaluating)

Evidence of Learning

- Graphic organizer "Inquiry Frayer Model"

- Graphic organizer "The Spheres of Science and Engineering Activity"

At a Glance

In this activity, participants will deepen their understanding of science inquiry by exploring the connections between the *Framework*'s spheres of activity for scientists and engineers and the science and engineering practices. The activity assumes that participants have already completed one of the two model activities in Activity 5. If you are not using this activity in conjunction with Activity 5, you will need to add an appropriate model inquiry activity.

Facilitator's Notes

Over at least the past two decades, the cornerstone of good science teaching has been the use of inquiry. However, we have struggled to define inquiry in a consistent manner that remains authentic to the scientific endeavor. Too often, the process of science is portrayed as a linear set of steps—or a constrained experimental method—that does not reflect actual science. The *Framework* attempts to move our definition of inquiry forward by defining a set of practices that represent the ways in which scientists and engineers engage in their work. The *Framework* authors write:

One helpful way of understanding the practices of scientists and engineers is to frame them as work that is done in three spheres of activity. … In one sphere, the dominant activity is investigation and empirical inquiry. In the second, the essence of work is the construction of explanations or designs using reasoning, creative thinking, and models. And in the third sphere, the ideas, such as the fit of models and explanations to evidence or the appropriateness of product designs, are analyzed, debated, and evaluated. In all three spheres of activity, scientists and engineers try to use the best available tools to support the task at hand, which today means that modern computational technology is integral to virtually all aspects of their work. (pp. 44–45)

By thinking about the work of scientists and engineers in three spheres, we can consider the wide range of practices that are used. The *Framework* states:

The focus here is on important practices, such as modeling, developing explanations, and engaging in critique and evaluation (argumentation), that have too often been underemphasized in the context of science education. In particular, we stress that critique is an essential element both for building new knowledge in general and for the learning of science in particular. Traditionally, K–12 science education has paid little attention to the role of critique in science. However, as all ideas in science are evaluated against alternative explanations and compared with evidence, acceptance of an explanation is ultimately an assessment of what data are reliable and relevant and a decision about which explanation is the most satisfactory. Thus knowing why the wrong answer is wrong can help secure a deeper and stronger understanding of why the right answer is right. Engaging in argumentation from evidence about an explanation supports students' understanding of the reasons and empirical evidence for that explanation, demonstrating that science is a body of knowledge rooted in evidence. (p. 44)

FIGURE 6.1

The work of scientists in the three spheres

Investigating

In this sphere, scientists ask questions, make observations, and collect data using a variety of methods.

Evaluating

In this sphere, scientists critique and analyze explanations and models. They engage in argumentation to test claims based on existing evidence and propose alternative explanations.

Explaining/ Solving

In this sphere, scientists attempt to make sense of phenomena by using evidence to create or modify explanations and models. They draw from established theories and models and often propose hypotheses that can be tested.

For more details on the practices of science and engineering and these three spheres of activity, read pages 43–48 of the *Framework*. The work of scientists in these three spheres is summarized in Figure 6.1 (see matching handout on p. 57).

When evaluating a science inquiry activity for classroom use, consider the questions listed below (adapted from NRC 2000). (These questions also provide guidance when determining how to modify classroom investigations.)

- Do students identify their own scientifically oriented question or is a question given to them?

- Do students have an opportunity to determine what data should be collected and to design methods to collect that data?

- Does the activity encourage students to seek out additional information, connect to known science, or share results with classmates?

- Does the activity engage students in making sense of data by generating an evidence-based explanation to attempt to answer the question?

- Does the activity provide an opportunity for students to engage in argumentation from evidence by justifying their explanations or critiquing the explanations of others?

Materials

- Copies of the following handouts: "Inquiry Frayer Model," (p. 54) and "The Spheres of Science and Engineering Activity" (p. 55)

- Access to the "Science and Engineering Practices" matrix (*NGSS*, Appendix F) or the "Science and Engineering Practices Self-Assessments" handout from Activity 8 (pp. 70–80)

Procedure

Set-up: Organize participants into small groups prior to starting this activity. When possible, participants should be in mixed grade-level groups.

Introduction (10 minutes): Introduce this activity by asking participants to complete the "Inquiry Frayer Model." This can be completed individually or in pairs. If you would prefer to make participants' initial knowledge more visible, have participants complete this Frayer model on chart paper instead of individual handouts. A Frayer model is an excellent graphic organizer for helping learners better understand important vocabulary. It consists of four quadrants with specific prompts (definition, characteristics, examples, nonexamples) related to the central term or phrase (in this case, inquiry). Participants should share their initial Frayer models with others before moving on to the next part.

Science Inquiry and the Spheres of Science and Engineering Activity (40 minutes): Introduce the spheres of activity for scientists and engineers (see Facilitator's Notes). Give participants the handout, "The Spheres of Science and Engineering Activity," and instruct groups to describe how the model activity engaged participants in the three spheres. Participants should also critique the activity based on the questions included in the handout. Briefly discuss their thoughts as a group.

Science Inquiry and the Science and Engineering Practices (20 minutes): Ask participants to reflect on the science and engineering practices (using either the *NGSS* matrix or the self-assessments in Activity 8). In this part of the activity, participants identify multiple practices that were used in the model activity. Working in small groups, they select one or two of the practices and determine how they can modify the model activity to further support students' ability to use those practices. In addition, participants identify and describe ways they could assess those practices. Provide time for participants to share their activity modifications and assessment ideas with the entire group.

Debrief (10 minutes): Participants should return to their initial Frayer model and make revisions as necessary. After providing time to make these revisions, have participants share any major changes that they made by conducting a quick "whip" around the room.

ACTIVITY 6
Inquiry Frayer Model

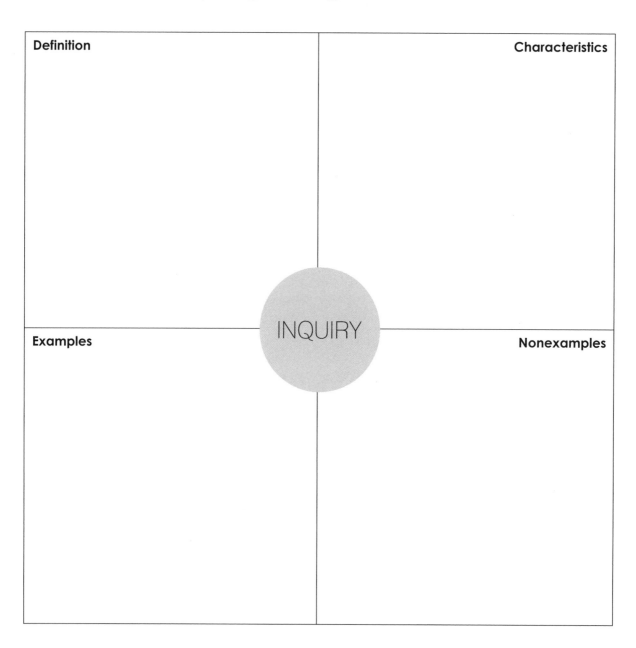

Definition	Characteristics

INQUIRY

Examples	Nonexamples

NATIONAL SCIENCE TEACHERS ASSOCIATION

ACTIVITY 6

The Spheres of Science and Engineering Activity

The *Framework* describes how engineers engage in the practices within these three spheres of activity. How were these spheres illustrated in the model activity?

Investigating	Evaluating	Explaining/Solving
In this sphere, scientists ask questions, make observations, and collect data using a variety of methods. How did you engage in this sphere during the model activity?	*In this sphere, scientists critique and analyze explanations and models. They engage in argumentation to test claims based on existing evidence and propose alternative explanations.* How did you engage in this sphere during the model activity?	*In this sphere, scientists attempt to make sense of phenomena by using evidence to create or modify explanations and models. They draw from established theory and models and often propose hypotheses that can be tested.* How did you engage in this sphere during the model activity?

ACTIVITY 6

The Spheres of Science and Engineering Activity

When evaluating a science inquiry activity for classroom use, consider the questions listed below (adapted from NRC 2000). (These questions also provide guidance when determining how to modify classroom investigations.)

- Do students identify their own scientifically oriented question or is a question given to them?

- Do students have an opportunity to determine what data should be collected and design methods to collect that data?

- Does the activity encourage students to seek out additional information, connect to known science, and/or share results with classmates?

- Does the activity engage students in making sense of data by generating an evidence-based explanation to attempt to answer the question?

- Does the activity provide an opportunity for students to engage in argumentation from evidence—justifying their explanations and/or critiquing the explanations of others?

Use these questions to critique the model activity.

Which science and engineering practice(s) do you want to emphasize in this activity?	How will you teach or assess this practice?

ACTIVITY 6

The Work of Scientists in the Three Spheres

Investigating

In this sphere, scientists ask questions, make observations, and collect data using a variety of methods.

Evaluating

In this sphere, scientists critique and analyze explanations and models. They engage in argumentation to test claims based on existing evidence and propose alternative explanations.

Explaining/ Solving

In this sphere, scientists attempt to make sense of phenomena by using evidence to create or modify explanations and models. They draw from established theories and models and often propose hypotheses that can be tested.

ACTIVITY 7

Engineering Design and the Practices of Science and Engineering

With Reynee Kachur

Approximate Length

110–120 minutes

Objectives

During this activity, participants will

- explore a model engineering design activity,

- develop an understanding of the engineering design process,

- determine at least two different possible solutions to an engineering design problem,

- demonstrate with a written response the connection(s) between science content and the model engineering design activity,

- describe how the model engineering design activity engaged participants in the three spheres, and

- critique the model engineering design activity.

Vocabulary

- three spheres (investigating, explaining/solving, evaluating)

- engineering design process

- science and engineering practices

Evidence of Learning

- Graphic organizer "The Spheres of Science and Engineering Activity"

- Written response demonstrating connections

- Solutions to the engineering design problem

At a Glance

In this activity, participants engage in a model activity as they develop an understanding of the engineering design process (as defined in the *NGSS*) and the science and engineering practices. Participants should have a basic understanding of the structure of the *NGSS* before starting in this activity. It will also be helpful if participants have some familiarity with the *Framework*'s spheres of activity for scientists and engineers (see Activity 6, pp. 50–56).

Facilitator's Notes

Before using this activity, you should be familiar with the information in *NGSS* Appendix I, "Engineering Design." You should also be familiar with the *Framework*'s "Spheres of Science and Engineering Activity" as described in Activity 6, pages 50–56. This activity focuses on these spheres from an engineering perspective.

The *Framework* and the *NGSS* define engineering design as having three components:

1. Defining and delimiting engineering problems involves stating the problem to be solved as clearly as possible in terms of criteria for success, and constraints or limits.

2. Designing solutions to engineering problems begins with generating a number of different possible solutions, then evaluating potential solutions to see which ones best meet the criteria and constraints of the problem.

3. Optimizing the design solution involves a process in which solutions are systematically tested and refined and the final design is improved by trading off less important features for those that are more important. (*NGSS*, Appendix I, p. 2)

In addition to these three components, we feel that it is important to add two more. First, in order to be effective learning activities in a science unit, engineering design activities must have strong connections to the science content of the unit, and these connections must be made explicit to the students. Second, engineering design is a social process. Students should be expected to communicate and justify their proposed solutions.

Particular emphasis needs to be given to two aspects of components B and C. Early in the design process, learners should brainstorm multiple ways to solve the problem. Quite often, learners new to engineering design settle quickly on their first potential solution instead of brainstorming and analyzing a variety of ideas. It is also important to encourage learners to build, test, and revise multiple prototypes. In order to implement engineering design in an authentic manner, sufficient time needs to be provided for this iterative process.

Engineering Design and the Science and Engineering Practices

The connection between engineering design activities and the science and engineering practices is also important. The practices clearly describe what students should be able to do as they approach problems and learn new science content. The grade band statements located in the foundations boxes and in the "Science and Engineering Practices" matrix (*NGSS*, Appendix F) should guide instruction as teachers build capacity for students

to engage in the practices and also provide measurable outcomes that can be assessed.

The *Framework* describes three spheres (investigating, evaluating, explaining/solving) of activity that define the work of scientists and engineers. These spheres were introduced in Activity 6. If you have not completed Activity 6 (pp. 50–56) with the participants you are working with now, you will need to introduce the importance of thinking about classroom activities through the lens of these spheres.

The work of engineers within these three spheres is summarized in Figure 6.2 (p. 60; see matching handout on p. 65).

When evaluating a science inquiry activity for classroom use, consider the questions listed below. (These questions also provide guidance when determining how to modify engineering activities.)

- Is the design activity connected to appropriate science content? Does the activity build explicit connections to that science content?

- Are students engaged in defining a problem?

- Are students encouraged to brainstorm multiple possible solutions?

- Do students have the opportunity to test and improve designs?

- Does the activity provide an opportunity for students to engage in argumentation from evidence—justifying their design decisions and/or critiquing the solutions of others?

Materials

- Copies of the following handouts: "The Spheres of Science and Engineering Activity" (pp. 62–63) and "Science Investigations Versus Engineering Design," (p. 64)

- Access to the "Science and Engineering Practices" matrix (*NGSS*, Appendix F) or copies of the "Science and Engineering Practices Self-Assessments" handout from Activity 8 (pp. 70–80).

- A video that shows engineering design in action.

  ○ The ABC Nightline episode "Deep Dive" (broadcast 07/13/99) focused on a challenge to build a better shopping cart. It provides an excellent introduction to the design process. The video can be ordered here: *http://films.com/id/11160/The_Deep_Dive_One_Companys_Secret_Weapon_for_Innovation.htm.*

  ○ Many alternatives exist for the "Deep Dive" video. PBS Design Squad has a number of videos that are suitable. We have used "Designing Swimming Prosthetics for a Dancer" (6 minutes: *http://pbskids.org/designsquad/parentseducators/guides/adaptive_technologies.html*) and "Balloon Joust" (2.5 minutes: *http://pbskids.org/designsquad/video/?guid=e30758ea-3576-48d5-8523-fb043e5a3b3f*).

- Model engineering design activity: Participants should actively experience the engineering design process. Example activities can be found at the PBS Design Squad website. We have successfully used the "Zip Line" and "Touchdown" activities from the NASA/Design Squad "On the Moon" educator guide in our workshops: *http://pbskids.org/designsquad/parentseducators/guides/index.html.*

FIGURE 6.2

The work of engineers in the three spheres

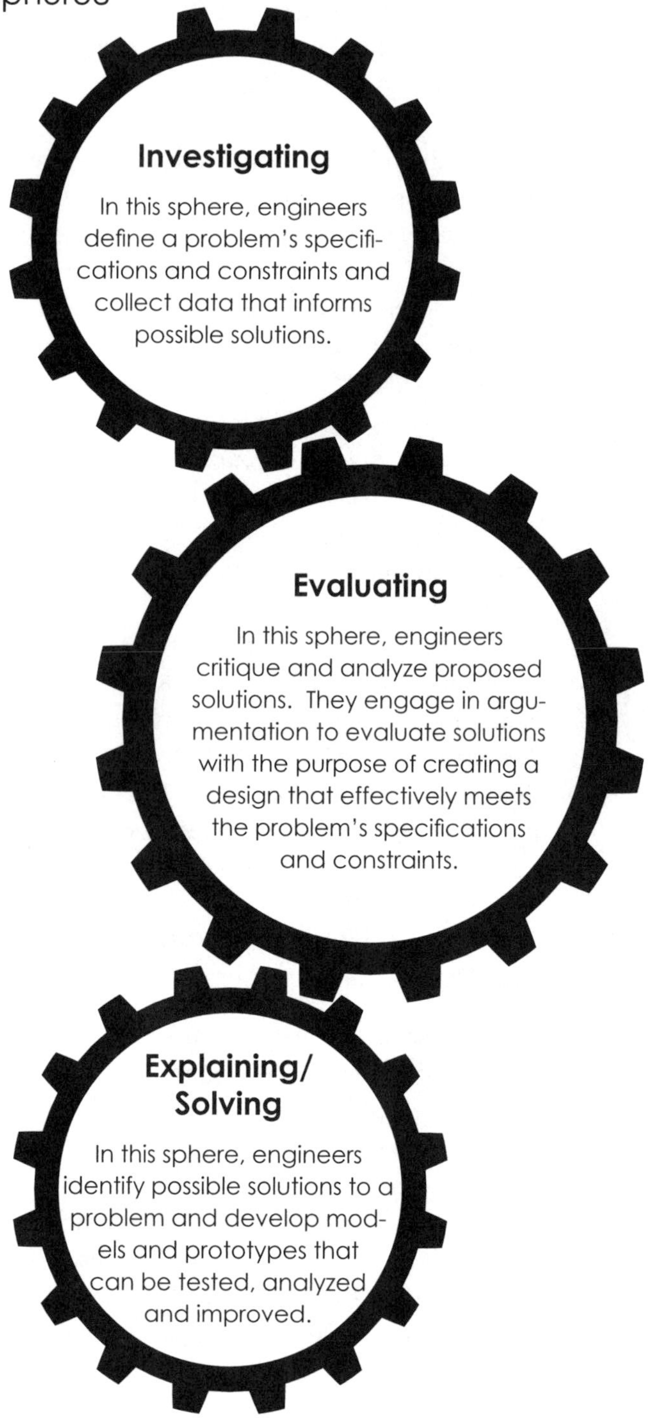

Investigating

In this sphere, engineers define a problem's specifications and constraints and collect data that informs possible solutions.

Evaluating

In this sphere, engineers critique and analyze proposed solutions. They engage in argumentation to evaluate solutions with the purpose of creating a design that effectively meets the problem's specifications and constraints.

Explaining/ Solving

In this sphere, engineers identify possible solutions to a problem and develop models and prototypes that can be tested, analyzed and improved.

Procedure

Set-up: Organize participants into small groups prior to starting this activity. When possible, participants should be in mixed grade-level groups.

Introduction (5 minutes + video): Begin by explaining that the purpose of this activity is to develop an understanding of the engineering design process as defined by the *NGSS*. Show the video. Afterward, ask participants to brainstorm characteristics of an engineering design process.

Engineering Design Activity Part I (40 minutes): Introduce the model engineering design activity. In addition to the structure provided by the activity, we find that it is necessary to require children and adults to determine at least two possible solutions before they are allowed to begin building. Many teachers new to engineering design will focus on the first solution that they identify—even if it is not plausible! Furthermore, it is critical that you help groups manage time effectively so that they have the opportunity to conduct multiple tests and revisions of their proposed solutions. It is common for novices to focus on building for the entire allotted time and not test until the very end. Finally, provide time at the end of the activity for participants to share their designs.

Engineering Design Activity Part II (10 minutes): If engineering design activities are being used in the science classroom, it is important that the connection between the activity and the content of the unit is made explicit. Provide participants with a writing prompt that connects the model activity to science content. For example, if you use the "Zip Line" activity, you may ask participants to describe how their zip line illustrates Newton's first and third laws. After giving participants time to write, ask them to share their responses.

Features of Engineering Design (20 minutes): Introduce the spheres of activity for scientists and engineers from an engineering perspective (see Facilitator's Notes). Give participants the handout for "The Spheres of Science and Engineering Activity" and instruct groups to describe how the model activity engaged participants in the three spheres. Participants should also critique the activity based on the questions included in the handout. Briefly discuss their thoughts as a group.

Engineering Design and the Science and Engineering Practices (30 minutes): Ask participants to reflect on the science and engineering practices (using either the *NGSS* matrix or the self-assessments in Activity 8. In this part of the activity, participants identify multiple practices used in the model activity. Working in small groups, they select one or two of the practices and determine how they can modify the model activity to further support students' ability to use those practices. In addition, participants should identify and describe ways that they could assess those practices. Participants should share their activity modifications and assessment ideas with the entire group.

Debrief (15 minutes): Provide participants with a copy of the "Science Investigations vs. Engineering Design" handout. Participants should complete the Venn diagram in small groups. After completing the diagram, have participants report out by having each group share one similarity and one difference between the two types of activities.

Next Steps

The NSTA webinar, "Engineering Practices in the *NGSS*" (*http://learningcenter.nsta.org/products/symposia_seminars/NGSS/webseminar15.aspx*) is an excellent next step for teachers interested in learning more about how engineering design is integrated into *NGSS*.

ACTIVITY 7

The Spheres of Science and Engineering Activity

The *Framework* describes how engineers engage in the practices within these three spheres of activity. How were these spheres illustrated in the model activity?

Investigating	Evaluating	Explaining / Solving
In this sphere, engineers define a problem's specifications and constraints and collect data that informs possible solutions. How did you engage in this sphere during the model activity?	*In this sphere, engineers critique and analyze proposed solutions. They engage in argumentation to evaluate solutions with the purpose of creating a design that effectively meets the problem's specifications and constraints.* How did you engage in this sphere during the model activity?	*In this sphere, engineers identify possible solutions to a problem and develop models and prototypes that can be tested, analyzed and improved.* How did you engage in this sphere during the model activity?

ACTIVITY 7

The Spheres of Science and Engineering Activity

The *NGSS* and the *Framework* define three components of engineering design: defining a problem, identifying solutions, and optimizing the solution. A good engineering design activity should also connect to science content and provide opportunities for students to justify and critique proposed solutions.

When evaluating a science inquiry activity for classroom use, consider the questions listed below. (These questions also provide guidance when determining how to modify engineering activities.)

- Is the design activity connected to appropriate science content? Does the activity build explicit connections to that science content?

- Are students engaged in defining a problem?

- Are students encouraged to brainstorm multiple possible solutions?

- Do students have the opportunity to test and improve designs?

- Does the activity provide an opportunity for students to engage in argumentation from evidence—justifying their design decisions or critiquing the solutions of others?

Use these questions to critique the model activity.

Which science and engineering practice(s) do you want to emphasize in this activity?	How will you teach or assess this practice?

ACTIVITY 7

Science Investigations Versus Engineering Design

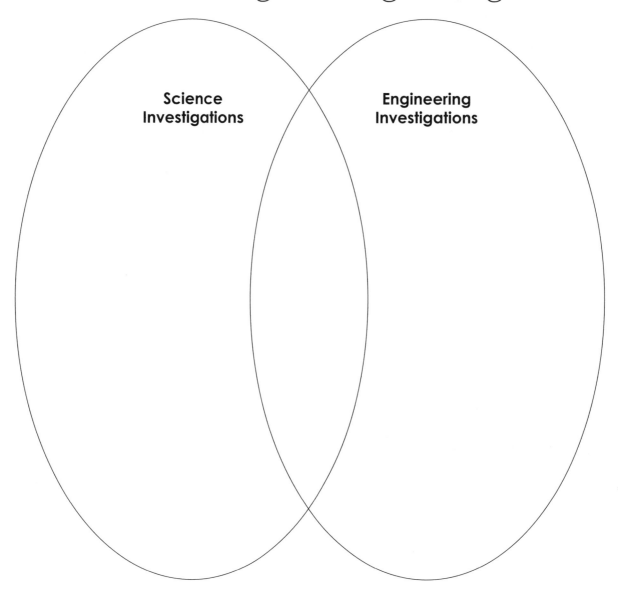

Science
Investigations

Engineering
Investigations

ACTIVITY 7

The Work of Engineers in the Three Spheres

Investigating

In this sphere, engineers define a problem's specifications and constraints and collect data that informs possible solutions.

Evaluating

In this sphere, engineers critique and analyze proposed solutions. They engage in argumentation to evaluate solutions with the purpose of creating a design that effectively meets the problem's specifications and constraints.

Explaining/ Solving

In this sphere, engineers identify possible solutions to a problem and develop models and prototypes that can be tested, analyzed and improved.

ACTIVITY 8

Science and Engineering Practices Self-Assessment

Approximate Length

30–40 minutes

Objectives

During this activity, participants will

- self-assess their understanding of the science and engineering practices.

Vocabulary

- science and engineering practices

Evidence of Learning

- Chart of group data

At a Glance

In this activity, participants self-assess their understanding of the science and engineering practices. After the self-assessment, group data is collected related to the practices that participants are most and least comfortable with.

Facilitator's Notes

The *Framework* identified a small number of disciplinary core ideas that all students should learn with increasing depth and sophistication, from kindergarten through grade 12. Key to the vision expressed in the *Framework* is for students to learn these disciplinary core ideas in the context of science and engineering practices. The importance of combining science and engineering practices and disciplinary core ideas is stated in the *Framework* as follows:

Standards and performance expectations that are aligned with the Framework must take into account that students cannot fully understand scientific and engineering ideas without engaging in the practices of inquiry and the discourses by which such ideas are developed and refined. At the same time, they cannot learn or show competence in practices except in the context of specific content. (p. 218)

The *Framework* specifies that each performance expectation must combine a relevant practice of science or engineering with a core disciplinary idea and crosscutting concept appropriate for students of the designated grade level. That guideline is perhaps the most significant way in which the *NGSS* differ from prior standards. In the future, science assessments will not assess students' understanding of core ideas separately from their abilities to use the practices of science and engineering. They will be assessed together, showing that students not only "know" science concepts but can also use their understanding to investigate the natural world through the practices of science inquiry, or solve meaningful problems through the practices of engineering design. The *Framework* uses the term "practices," rather than "science processes" or "inquiry" skills for a specific reason: "We use the term 'practices' instead of a term such as 'skills' to emphasize that engaging in scientific investigation requires not only skill but also knowledge that is specific to each practice" (NRC 2012, p. 30).

Materials

- Appropriate "Science and Engineering Practices Self-Assessments" (pp. 70–80) for each participant. The self-assessments are located at the end of this activity.

- (Optional) Copies of the "Modeling, Evidence, Explanations, and Argumentation" handout (p. 69)

- A poster or chart paper with the titles for each science and engineering practice

- Three green and three red sticker dots for each participant. As an alternative, participants can use green and red markers and draw an X or checkmark on the poster or chart paper.

Procedure

Introduction (5 minutes): Briefly introduce participants to the importance of the science and engineering practices by using an excerpt from Appendix F of the *NGSS*. (Excerpt located in Facilitator's Notes.)

Activating Prior Knowledge (Optional) (10 minutes): Although all eight science and engineering practices are important for student learning, three of the practices stand out as *primus inter pares*—first among equals. Practice 2 (Develop and Use Models), the "science" component of practice 6 (Constructing Explanations), and practice 7 (Engaging in Argument From Evidence) are central to knowledge creation in science and will likely require significant attention during professional development and *NGSS* implementation. Attention can be drawn to these three practices by having participants reflect, in writing, on the following prompts prior to completing the

self-assessment. Participants return to their reflection at the end of the activity. These prompts are also located in the "Models, Evidence, Explanations, and Argumentation" handout.

Prompts:

- How do scientists use models to help them understand the natural world?

- How do (or can) you engage students in the use of models in your course?

- How do scientists use evidence to create explanations and engage in argumentation?

- How do (or can) you engage students in using evidence to create explanations and in argumentation?

Science and Engineering Self-Assessment (15–25 minutes): Provide each participant with an appropriate self-assessment. Participants should complete the self-assessment individually.

Debrief (10 minutes): The purpose of this debrief is to collect group data related to the practices that participants are most and least comfortable with. Before beginning data collection, reinforce that teacher understanding of the practices is on a continuum; a number of these practices should feel familiar to many participants. However, it is important to realize that we can all improve our understanding of how to build students' capacity to understand and use each practice. Provide each participant with three red and three green sticker dots or red and green markers. Participants should use red to identify the three practices that they are the least comfortable with and green to identify the practices that they are most comfortable with. You may want to conclude this session by providing participants with a link to

NSTA's webinar series on the science and engineering practices (see the Next Steps section).

Debrief 2 (Optional) (10–20 minutes): Ask participants to make revisions to their initial thoughts about modeling, explanations, and argumentation. It may be helpful to refer back to the grade-band descriptors provided in the self-assessment documents. Have participants share their top one or two insights related to these three practices.

Next Steps

The purpose of this activity is to collect data about participants' comfort and expertise with the standards. This data can be used to plan future professional development. Encourage participants to read the chapter on science and engineering practices from *A Framework for K–12 Science Education* (NRC 2012) to deepen their understanding of the practices. In addition, participants can be referred to the NSTA webinar series on the science and engineering practices as a way to deepen their understanding of specific practices (*http://learningcenter.nsta.org/products/web_seminar_archive_sponsor.aspx?page=NGSS*).

ACTIVITY 8

Modeling, Evidence, Explanations, and Argumentation

Modeling	Use of evidence: Explanations and argumentation
How do scientists use models to help them understand the natural world?	How do scientists use evidence to create explanations and engage in argumentation?
How do (or can) you engage students in the use of models in your course?	How do (or can) you engage students in using evidence to create explanations and in argumentation?

ACTIVITY 8

Science and Engineering Practices Self-Assessments

K—2

NGSS Science and Engineering Practices, Grades K–2 1 = Little Understanding; 4 = Expert Understanding					
Practice / Indicator	**1**	**2**	**3**	**4**	**NOTES**
Asking questions and defining problems in grades K–2 builds on prior experiences and progresses to simple descriptive questions that can be tested.					
Ask questions about observations of the natural and designed world.					
Modeling in K–2 builds on prior experiences and progresses to include identifying, using, and developing models that represent concrete events or design solutions.					
Distinguish between a model and the actual object, process, and/or events the model represents.					
Compare models to identify common features and differences.					
Develop and/or use models (i.e., diagrams, drawings, physical replicas) that represent amounts, relative scales (bigger, smaller), and patterns.					
Planning and carrying out investigations to answer questions or test solutions to problems in K–2 builds on prior experiences and progresses to simple investigations, based on fair tests, which provide data to support explanations or design solutions. *Planning and carrying out investigations may include elements of all of the other practices.*					
Plan and carry out investigations collaboratively.					
Evaluate different ways of observing and/or measuring an attribute of interest.					
Make observations and/or measurements to collect data that can be used to make comparisons.					
Identify questions and make predictions based on prior experiences.					

NGSS Science and Engineering Practices, Grades K–2 1 = Little Understanding; 4 = Expert Understanding					
Practice / Indicator	**1**	**2**	**3**	**4**	**NOTES**
Analyzing data in K–2 builds on prior experiences and progresses to collecting, recording, and sharing observations.					
Use and share pictures, drawings, and/or writings of observations where appropriate.					
Use observations to note patterns and/or relationships in order to answer scientific questions and solve problems.					
Make measurements of length using standard units to quantify data.					
Mathematical and computational thinking at the K–2 level builds on prior experience and progresses to recognizing that mathematics can be used to describe the natural and designed world.					
Decide when to use qualitative vs. quantitative data.					
Use data to identify patterns in the natural and designed worlds.					
Use standard units to measure and compare the lengths of different objects and display the data using simple graphs.					
Constructing explanations and designing solutions in K–2 builds on prior experiences and progresses to the use of evidence or ideas in constructing explanations and designing solutions.					
Use information from observations to construct explanations about investigations.					
Use tools and materials provided to design a solution to a specific problem.					
Distinguish between opinions and evidence.					
Engaging in argument from evidence in K–2 builds on prior experiences and progresses to comparing ideas and representations about the natural and designed world.					
Distinguish arguments that are supported by evidence from those that are not.					
Listen actively to others' arguments and ask questions for clarification.					
Obtaining, evaluating, and communicating information in K–2 builds on prior experiences and uses observations and texts to communicate new information.					
Read and comprehend grade appropriate texts and/or use other reliable media to acquire scientific and/or technical information.					
Critique and communicate information or design ideas with others in oral and/or written forms using models, drawings, writing, or numbers.					
Record observations, thoughts, and ideas.					

3–5

NGSS Science and Engineering Practices, Grades 3–5 1 = Little Understanding; 4 = Expert Understanding					
Practice / Indicator	**1**	**2**	**3**	**4**	**NOTES**
Asking questions and defining problems in grades 3–5 builds from grades K–2 experiences and progresses to specifying qualitative relationships.					
Identify scientific (testable) and nonscientific questions.					
Ask questions based on careful observations of phenomena and information.					
Ask questions of others to clarify ideas or request evidence.					
Ask questions that relate one variable to another variable.					
Ask questions to clarify the constraints of solutions to a problem					
Modeling in 3–5 builds on K–2 models and progresses to building and revising simple models and using models to represent events and design solutions.					
Construct and revise models collaboratively to measure and explain frequent and regular events.					
Construct a model using an analogy, example, or abstract representation to explain a scientific principle or design solution.					
Use simple models to describe phenomena and test cause and effect relationships concerning the functioning of a natural or designed system.					
Identify limitations of models.					
Planning and carrying out investigations to answer questions or test solutions to problems in 3–5 builds on K–2 experiences and progresses to include investigations that control variables and provide evidence to support explanations or design solutions. Planning and carrying out investigations may include elements of all of the other practices.					
Plan and carry out investigations collaboratively, using fair tests in which variables are controlled and the number of trials considered.					
Discuss and evaluate appropriate methods and tools for collecting data.					
Make observations and/or measurements, collect appropriate data, and identify patterns that provide evidence to explain a phenomenon or test a design solution.					
Formulate questions and predict reasonable outcomes based on patterns such as cause and effect relationships.					

NGSS Science and Engineering Practices, Grades 3–5 1 = Little Understanding; 4 = Expert Understanding					
Practice / Indicator	**1**	**2**	**3**	**4**	**NOTES**
Analyzing data in 3–5 builds on K–2 and progresses to introducing quantitative approaches to collecting data and conducting multiple trials of qualitative observations.					
Display data in tables and graphs, using digital tools when feasible, to reveal patterns that indicate relationships.					
Use data to evaluate claims about cause and effect.					
Compare data collected by different groups in order to discuss similarities and differences in their findings.					
Use data to evaluate and refine design solutions.					
Mathematical and computational thinking at the 3–5 level builds on K–2 and progresses to extending quantitative measurements to a variety of physical properties and using computation and mathematics to analyze data and compare alternative design solutions.					
Use mathematical thinking and/or computational outcomes to compare alternative solutions to an engineering problem.					
Analyze simple data sets for patterns that suggest relationships.					
Use standard units to measure area, volume, weight, and temperature.					
Constructing explanations and designing solutions in 3–5 builds on prior experiences in K–2 and progresses to the use of evidence in constructing multiple explanations and designing multiple solutions.					
Use quantitative relationships to construct explanations of observed events (e.g., the distribution of plants in the back yard or why some things sink and others float).					
Use evidence (e.g., measurements, observations, patterns) to construct a scientific explanation or solution to a problem.					
Identify the evidence that supports an explanation.					
Apply scientific knowledge to solve design problems.					
Engaging in argument from evidence in 3–5 builds from K–2 experiences and progresses to critiquing the scientific explanations or solutions proposed by peers by citing relevant evidence about the natural and designed world.					
Construct and/or support scientific arguments drawing on evidence, data, or a model.					
Compare and refine multiple arguments based on the strengths and weaknesses of the evidence supporting the argument.					
Respectfully provide and receive critique on the scientific arguments proposed by peers by citing relevant evidence and/or making logical arguments.					

NGSS Science and Engineering Practices, Grades 3–5 1 = Little Understanding; 4 = Expert Understanding					
Practice / Indicator	**1**	**2**	**3**	**4**	**NOTES**
Obtaining, evaluating, and communicating information in 3–5 builds on K–2 and progresses to evaluating the merit and accuracy of ideas and methods.					
Compare and/or synthesize across texts and/or other reliable media to acquire and/or generate appropriate scientific and/or technical information.					
Synthesize information in written text with that contained in corresponding tables, diagrams, and/or charts.					
Generate and communicate scientific and/or technical information orally and/or in written formats using various forms of media and may include tables, diagrams, and charts.					
Use models to share findings or solutions in oral and/or written presentations, and/or extended discussions.					

6—8

NGSS Science and Engineering Practices, Grades 6–8 1 = Little Understanding; 4 = Expert Understanding					
Practice / Indicator	**1**	**2**	**3**	**4**	**NOTES**
Asking questions and defining problems in grades 6–8 builds from grades K–5 experiences and progresses to formulating and refining empirically testable questions and explanatory models					
Ask questions that arise from phenomena, models, or unexpected results.					
Ask questions to clarify or identify the premise(s) of an argument.					
Ask questions to determine relationships between independent and dependent variables.					
Ask questions that challenge the interpretation of a data set.					
Ask questions to refine a model, an explanation, or an engineering problem.					
Modeling in 6–8 builds on K–5 and progresses to developing, using, and revising models to explain, explore, and predict more abstract phenomena and design systems.					
Use and/or construct models to predict, explain, and/or collect data to test ideas about phenomena in natural or designed systems, including those representing inputs and outputs.					
Pose models to describe mechanisms at unobservable scales.					
Modify models—based on their limitations—to increase detail or clarity, or to explore what will happen if a component is changed.					
Use and construct models of simple systems with uncertain and less predictable factors.					
Planning and carrying out investigations to answer questions or test solutions to problems in 6–8 builds on K–5 experiences and progresses to include investigations that use multiple variables and provide evidence to support explanations or design solutions.					
Plan and carry out investigations individually and collaboratively, identifying independent and dependent variables and controls.					
Discuss and evaluate the accuracy of various methods for collecting data.					
Collect data and generate evidence to answer scientific questions or test design solutions under a range of conditions.					

NGSS Science and Engineering Practices, Grades 6–8 1 = Little Understanding; 4 = Expert Understanding					
Practice / Indicator	**1**	**2**	**3**	**4**	**NOTES**
Formulate a question that can be investigated within the scope of the classroom, school laboratory, or field with available resources and, when appropriate, frame a hypothesis (a possible explanation that predicts a particular and stable outcome) based on a model or theory					
Analyzing data in 6–8 builds on K–5 and progresses to extending quantitative analysis to investigations, distinguishing between correlation and causation, and basic statistical techniques of data and error analysis.					
Use mean, median, mode, and variability to analyze and characterize data.					
Use graphical displays to analyze data in order to identify linear and nonlinear relationships.					
Consider limitations of data analysis, such as measurement error, and seek to improve precision and accuracy of data with better technological tools and methods such as multiple trials.					
Distinguish between causal and correlational relationships.					
Use data to define an operational range for a design solution.					
Use graphical displays (e.g., maps) of large data sets to identify temporal and spatial relationships.					
Mathematical and computational thinking at the 6–8 level builds on K–5 and progresses to identifying patterns in large data sets and using mathematical concepts to support explanations and arguments.					
Use digital tools (e.g., computers) to analyze very large data sets for patterns and trends.					
Use mathematical concepts such as ratios, averages, basic probability, and simple functions, including linear relationships, to analyze data.					
Use mathematical arguments to justify scientific conclusions and design solutions.					
Constructing explanations and designing solutions in 6–8 builds on K–5 experiences and progresses to include constructing explanations and designing solutions supported by multiple sources of evidence consistent with scientific knowledge, principles, and theories.					
Construct explanations for either qualitative or quantitative relationships between variables.					
Apply scientific reasoning to show why the data are adequate for the explanation or conclusion.					

NGSS Science and Engineering Practices, Grades 6–8 1 = Little Understanding; 4 = Expert Understanding					
Practice / Indicator	**1**	**2**	**3**	**4**	**NOTES**
Base explanations on evidence and the assumption that natural laws operate today as they did in the past and will continue to do so in the future.					
Undertake design projects, engaging in the design cycle, to construct and implement a solution that meets specific design criteria and constraints.					
Apply scientific knowledge to explain real-world examples or events and solve design problems.					
Construct explanations from models or representations.					
Engaging in argument from evidence in 6–8 builds from K–5 experiences and progresses to constructing a convincing argument that supports or refutes claims for either explanations or solutions about the natural and designed world.					
Use oral and written arguments supported by empirical evidence and reasoning to support or refute an argument for a phenomenon or a solution to a problem.					
Evaluate competing design solutions based on jointly developed and agreed-upon design criteria.					
Compare two arguments from evidence to identify which is better by identifying flaws in logic or methods.					
Obtaining, evaluating, and communicating information in 6–8 builds on K–5 and progresses to evaluating the merit and validity of ideas and methods.					
Communicate understanding of scientific information that is presented in different formats (e.g., verbally, graphically, textually, mathematically).					
Generate and communicate ideas using scientific language and reasoning.					
Gather, read, and explain information from appropriate sources and evaluate the credibility of the publication, authors, possible bias of the source, and methods used.					
Read critically using scientific knowledge and reasoning to evaluate data, hypotheses, conclusions, and competing information.					

NGSS Science and Engineering Practices, Grades 9–12 1 = Little Understanding; 4 = Expert Understanding					9–12
Practice / Indicator	**1**	**2**	**3**	**4**	**NOTES**
Asking questions and defining problems in grades 9–12 builds from grades K–8 experiences and progresses to formulating, refining, and evaluating empirically testable questions and explanatory models and simulations.					
Ask questions that arise from phenomena, models, theory, or unexpected results.					
Ask questions that require relevant empirical evidence.					
Ask questions to determine quantitative relationships between independent and dependent variables.					
Ask questions that challenge the premise of an argument, the interpretation of a data set, or the suitability of a design.					
Modeling in 9–12 builds on K–8 and progresses to using, synthesizing, and constructing models to predict and explain relationships between systems and their components in the natural and designed world.					
Use multiple types of models to represent and explain phenomena, and move flexibly between model types based on merits and limitations.					
Construct, revise, and use models to predict and explain relationships between systems and their components.					
Use models (including mathematical and computational) to generate data to explain and predict phenomena, analyze systems, and solve problems.					
Design a test of a model to ascertain its reliability.					
Examine merits and limitations of various models in order to select or revise a model that best fits the evidence or the design criteria.					
Planning and carrying out investigations to answer questions or test solutions to problems in 9–12 builds on K–8 experiences and progresses to include investigations that build, test, and revise conceptual, mathematical, physical, and empirical models. Planning and carrying out investigations may include elements of all of the other practices.					
Plan and carry out investigations individually and collaboratively and test designs as part of building and revising models, explaining phenomena, or testing solutions to problems. Consider possible confounding variables or effects and ensure the investigation's design has controlled for them.					
Evaluate various methods of collecting data (e.g., field study, experimental design, simulations) and analyze components of the design in terms of various aspects of the study. Decide types, how much, and accuracy of data needed to produce reliable measurement and consider any limitations on the precision of the data (e.g., number of trials, cost, risk, time).					

NGSS Science and Engineering Practices, Grades 9–12 1 = Little Understanding; 4 = Expert Understanding					
Practice / Indicator	**1**	**2**	**3**	**4**	**NOTES**
Select appropriate tools to collect, record, analyze, and evaluate data.					
Plan and carry out investigations and test design solutions in a safe and ethical manner including considerations of environmental, social, and personal impacts.					
Analyzing data in 9–12 builds on K–8 and progresses to introducing more detailed statistical analysis, the comparison of data sets for consistency, and the use of models to generate and analyze data.					
Use tools, technologies, and/or models (e.g., computational, mathematical) to generate and analyze data in order to make valid and reliable scientific claims or determine an optimal design solution.					
Consider limitations (e.g., measurement error, sample selection) when analyzing and interpreting data.					
Determine function fits to data, including slope, intercept, and correlation coefficient for linear fits.					
Compare and contrast various types of data sets (e.g., self-generated, archival) to examine consistency of measurements and observations.					
Evaluate the impact of new data on a working explanation of a phenomenon or design solution.					
Mathematical and computational thinking at the 9–12 level builds on K–8 and progresses to using algebraic thinking and analysis, a range of linear and nonlinear functions including trigonometric functions, exponentials and logarithms, and computational tools for statistical analysis to analyze, represent, and model data. Simple computational simulations are created and used based on mathematical models of basic assumptions.					
Use mathematical or algorithmic representations of phenomena or design solutions to create explanations, computational models, or simulations.					
Use mathematical expressions to represent phenomena or design solutions in order to solve algebraically for desired quantities.					
Use simple limit cases to test mathematical expressions, computer programs or algorithms, or simulations to see if a model "makes sense" by comparing the outcomes with what is known about the real world.					
Use statistical and mathematical techniques and structure data (e.g., displays, tables, graphs) to find regularities, patterns (e.g., fitting mathematical curves to data), and relationships in data.					

NGSS Science and Engineering Practices, Grades 9–12 1 = Little Understanding; 4 = Expert Understanding					
Practice / Indicator	**1**	**2**	**3**	**4**	**NOTES**
Constructing explanations and designing solutions in 9–12 builds on K–8 experiences and progresses to explanations and designs that are supported by multiple and independent student-generated sources of evidence consistent with scientific knowledge, principles, and theories.					
Make quantitative claims regarding the relationship between dependent and independent variables.					
Apply scientific reasoning, theory, and models to link evidence to claims and show why the data are adequate for the explanation or conclusion.					
Construct and revise explanations and arguments based on evidence obtained from a variety of sources (e.g., scientific principles, models, theories) and peer review.					
Base casual explanations on valid and reliable empirical evidence from multiple sources and the assumption that natural laws operate today as they did in the past and will continue to do so in the future.					
Apply scientific knowledge to solve design problems by taking into account possible unanticipated effects.					
Engaging in argument from evidence in 9–12 builds from K–8 experiences and progresses to using appropriate and sufficient evidence and scientific reasoning to defend and critique claims and explanations about the natural and designed world. Arguments may also come from current scientific or historical episodes in science.					
Criticize and evaluate arguments and design solutions in light of new evidence, limitations (e.g., trade-offs), constraints, and ethical issues.					
Evaluate the merits of competing arguments, design solutions, and/or models					
Evaluate the claims, evidence, and reasoning of currently accepted explanations or solutions as a basis for the merits of the arguments.					
Construct a counterargument that is based in data and evidence that challenges another proposed argument.					
Obtaining, evaluating, and communicating information in 9–12 builds on K–8 and progresses to evaluating the validity and reliability of the claims, methods, and designs.					
Critically read scientific literature adapted for classroom use to identify key ideas and major points and to evaluate the validity and reliability of the claims, methods, and designs.					
Generate, synthesize, communicate, and critique claims, methods, and designs that appear in scientific and technical texts or media reports.					
Recognize the major features of scientific and technical writing and speaking and produce written and illustrated texts or oral presentations that communicate ideas and accomplishments.					

NATIONAL SCIENCE TEACHERS ASSOCIATION

ACTIVITY 9

Exploring Crosscutting Concepts

With Emily Miller

Approximate Length

60–80 minutes

Objectives

During this activity, participants will

- explore the progressions in crosscutting concepts,

- identify key crosscutting concepts that are closely aligned with what they currently teach,

- examine the benefits of teaching with crosscutting concepts, and

- brainstorm instructional strategies for connecting course content and crosscutting concepts.

Vocabulary

- crosscutting concepts

- progressions

Evidence of Learning

- Completed progressions of crosscutting concepts

- Sharing of instructional strategies to make connections

At a Glance

Participants will explore the crosscutting concepts in *NGSS* during this activity. Participants begin by organizing statements from the "Crosscutting Concepts" matrix (*NGSS*, Appendix G) into grade band progressions, are then introduced to the importance of crosscutting concepts, and finally identify key crosscutting concepts that are strongly connected to what they currently teach.

Facilitator's Notes

As the term conveys, crosscutting concepts provide a powerful mechanism for understanding the connections between science disciplines. For example, the crosscutting concept of Energy and Matter: Flows, Cycles, and Conservation, describes the construct of energy and matter and how scientists use those constructs to understand concepts in every discipline—from understanding chemical reactions in the physical sciences, to weather patterns in the Earth sciences, and interactions in ecosystems in the life sciences. By repeated exposure to these crosscutting concepts throughout their K–12 science experiences, students will gain an appreciation for the connections between the core ideas and the practices of science and engineering. Crosscutting concepts introduce the shared language with which to talk about and build on experiences in the classroom. The *NGSS* writers state:

The Framework identifies seven crosscutting concepts that bridge disciplinary boundaries, uniting core ideas throughout the fields of science and engineering. Their purpose is to help students deepen their understanding of the disciplinary core ideas (pp. 2 and 8), and develop a coherent and scientifically based view of the world (p. 83). (NGSS, Appendix G, p. 1).

The writers also explain the importance of explicitly teaching the crosscutting concepts:

Although these ideas have been consistently included in previous standards documents the Framework recognizes that "students have often been expected to build such knowledge without any explicit instructional support. Hence the purpose of highlighting them as Dimension 2 of the Framework is to elevate their role in the development of standards, curricula, instruction, and assessments." (p. 83)

The writing team has continued this commitment by weaving crosscutting concepts into the performance expectations for all students—so they cannot be left out (*NGSS*, Appendix G, p. 2).

The crosscutting concepts also present a great opportunity to specifically address the achievement gap that exists with many student populations. The *NGSS* were written with "All Standards, All Students" in mind. Research has shown that given an equitable learning situation, students from diverse backgrounds can learn science through the practices of science and engineering.

While students of nondominant groups steadily increase in the nation's schools, the teaching profession continues to consist of mostly white, middle-class females (Jorgenson 2000). This mismatch calls for an increased sensitivity and awareness to the different experiences students bring to the classroom. All students are constantly shuffling and reorganizing schema to fit into existing paradigms in order to make sense of the world. No student's paradigms mirror each other's exactly, but because dominant

and nondominant students and teachers often draw from different linguistic, cultural, and social experiences, their distinct functioning hypotheses of the world can present roadblocks to student learning.

Crosscutting concepts provide a common framework on which to anchor new scientific understanding; once both dominant and nondominant students understand a crosscutting concept, they share a common perspective that allows them to learn from each other. In general, using overarching concepts integrated into cohesive and well-coordinated units of study proves more effective for English language learners (Short and Echevarria 1999) as well as other diverse learners (Duschl, Schweingruber, and Shouse 2007) than teaching isolated, disconnected, seemingly unrelated ideas and events. Dominant students often already possess experiences that more closely match the science taught in the schools. Thus, the links between ideas are implied, and learned implicitly. However, for nondominant students the underlying key understanding and scientific practices may not be learned without explicit teaching.

Crosscutting concepts provide direction for purposeful links between the big ideas of science and the discrete components and level the playing field for nondominant students. Finally, it has been demonstrated that nondominant students can learn to incorporate what they already observe in the world into the broad curricular themes, allowing students to use these overarching concepts to enhance their existing mastery and understandings and to draw parallels from old to new knowledge (Whitmore and Crowel 1994).

The crosscutting concepts in *NGSS* offer more opportunities for educators to explore connections across the curriculum. For example, students must learn how identifying patterns aids understanding across all four scientific domains, but they can also explore patterns in music, art, and mathematics and subsequently deepen their comprehension of the concept. Likewise, cause and effect can be used to explain phenomena in Earth science as well as examining character or plot development in literature. Reinforcing the crosscurricular connections between and among crossdisciplinary units enhances familiarity with language and facilitates comprehension that can be specifically helpful to English language learners (Kucer, Silva, and Delgado-Larocco 1995).

Materials

- "Crosscutting Concepts Matrix" from *NGSS* Appendix G. To prepare for this activity, the individual bullet point statements need to be cut out into individual sentence strips. Statements should remain organized by crosscutting concept.

Procedure

Set-up: Participants should be assigned to groups prior to the start of the activity. This activity works best with groups that represent mixed grade levels (e.g., a group of three with an elementary, a middle, and a high school representative). You should also determine if you will assign groups to a specific crosscutting concept or allow them to select. You should be familiar with *NGSS* Appendix G and the Facilitator's Notes in this activity prior to facilitating the activity.

Introduction (5 minutes): Begin the activity by briefly describing that crosscutting concepts are ideas in science (like using patterns) that stretch across the disciplines. Provide participants with a

set of statements focused on one crosscutting concept with the instructions to assemble the statements into a progression (statements for K–2, 3–5, 6–8, and 9–12).

Group Work (15 minutes): Groups should work together to place statements for their crosscutting concept into a development progression by grade band.

Small Group Discussion (15 minutes): Within their groups, participants should discuss what they noticed about the crosscutting concept progressions and connections to what they currently teach. Groups should identify possible benefits to helping students understand how individual course content is related to crosscutting concepts.

Importance of Crosscutting Concepts (25 minutes): Begin this phase of the activity by asking groups to share the big ideas from their discussions on the benefits of using crosscutting concepts. As this discussion unfolds, reinforce these ideas with information from the Facilitator's Notes section of this activity.

Transfer (15 minutes): Participants should return to their small-group discussions and brainstorm possible strategies for making explicit connections between course content and crosscutting concepts (e.g., crosscutting concept anchor charts, frequent quick writes, portfolios, essential questions, and so on).

Debrief (10 minutes): Use a round robin reporting structure to allow groups to share their ideas. In this structure, group spokespersons take turns sharing one idea per group until all ideas have been exhausted or time runs out.

Next Steps

This activity introduced educators to the cross-cutting concepts and began the discussion about how teachers use them to build connections between units within a school year and across courses throughout a student's K–12 science experience. This discussion is continued in Activity 18 (p. 140), in which educators explore how to use crosscutting concepts to develop essential questions to guide instruction.

ACTIVITY 10

Integrating the Nature of Science

Approximate Length

90 minutes

Objectives

During this activity, participants will

- explore a model science activity that demonstrates the nature of science,

- reflect on how the model activity demonstrates the scientific process,

- categorize statements from the "Nature of Science Matrix" (*NGSS*, Appendix H, p. 6), and

- develop anchor charts.

Vocabulary

- nature of science

Evidence of Learning

- Graphic organizer "The Nature of Science"

- Anchor charts

At a Glance

Participants engage in a model activity as they learn about how the nature of science is reflected in the *NGSS*. They think about their understanding of the nature of science, understand the nature of science expectations in *NGSS*, and create draft anchor charts to develop student understanding of the nature of science.

Facilitator's Notes

A familiarity with Appendix H is sufficient for facilitating this activity. We find the following two excerpts from *NGSS* to be particularly helpful:

When the three dimensions of the science standards are combined, one can ask what is central to the intersection of the scientific and engineering practices, disciplinary core ideas, and crosscutting concepts? Or, what is the relationship among the three basic elements of A Framework for K–12 Science Education? Humans have a need to know and understand the world around them. And they have the need to change their environment using technology in order to accommodate what they understand or desire. In some cases, the need to know originates in satisfying basic needs in the face of potential dangers. Sometimes it is a natural curiosity and, in other cases, the promise of a better, more comfortable life. Science is the pursuit of explanations of the natural world, and technology and engineering are means of accommodating human needs, intellectual curiosity and aspirations. (NGSS, Appendix H, p. 2)

The nature of the scientific explanations is an idea central to standards-based science programs. Beginning with the practices, core ideas, and crosscutting concepts, science teachers can progress to the regularities of laws, the importance of evidence, and the formulation of theories in science. With the addition of historical examples, the nature of scientific explanations assumes a human face and is recognized as an ever-changing enterprise." (NGSS, Appendix H, p. 8)

The nature of science statements in *NGSS* are embedded in the science and engineering

FIGURE 6.3

Example *NGSS* page showing the incorporation of the nature of science

MS-ESS2-3   Earth's Systems

Students who demonstrate understanding can:

MS-ESS2-3.   **Analyze and interpret data on the distribution of fossils and rocks, contintental shapes, and seafloor structures to provide evidence of the past plate motions.** [Clarification Statement: Examples of data include similarities of rock and fossil types on different continents, the shapes of the continents (including continental shelves), and the locations of ocean structures (such as ridges, fracture zones, and trenches).] [*Assessment Boundary: Paleomagnetic anomalies in oceanic and continental crust are not assessed.*]

The performance expectation above was developed using the following elements from the NRC document *A Framework for K-12 Science Education*:

Science and Engineering Practices	Disciplinary Core Ideas	Crosscutting Concepts
Analyzing and Interpreting Data Analyzing data in 6–8 builds on K–5 experiences and progresses to extending quantitative analysis to investigations, distinguishing between correlation and causation, and basic statistical techniques of data and error analysis. • Analyze and interpret data to provide evidence for phenomena. *Connections to Nature of Science* **Scientific Knowledge is Open to Revision in Light of New Evidence** • Science findings are frequently revised and/or reinterpreted based on new evidence.	**ESS1.C: The History of Planet Earth** • Tectonic processes continually generate new ocean sea floor at ridges and destroy old sea floor at trenches. *(HS.ESS1.C GBE),(secondary)* **ESS2.B: Plate Tectonics and Large-Scale System Interactions** • Maps of ancient land and water patterns, based on investigations of rocks and fossils, make clear how Earth's plates have moved great distances, collided, and spread apart.	**Patterns** • Patterns in rates of change and other numerical relationships can provide information about natural systems.

practices and crosscutting concepts foundation boxes. For example, the *NGSS* performance expectation MS-ESS2-3 (Figure 6.3) incorporates nature of science concepts in the science and engineering practices foundations box.

Materials

- Copies of the handout "The Nature of Science" (p. 88)

- Chart paper and markers

- Access to the "Nature of Science Matrix" in Appendix H, "Nature of Science in the *NGSS*"

- The model activity "The 23rd Annual Consortium of Geologists," found in Appendix 4 (p. 224)

Procedure

Set-up: Participants should be assigned to grade-level or grade band groups prior to the start of this activity. Prepare enough copies of the images in the model activity for each group.

Model Activity (30 minutes): Participants work in small groups to complete the model activity ("The 23rd Annual Consortium of Geologists," Appendix 4).

Reflection (10 minutes): After completing the model activity, ask participants to reflect on how the activity models the scientific process.

Discussion and Exploration (30 minutes): Ask participants to share their reflections. Use the Facilitator's Notes to explain that *NGSS* integrates the characteristics of the nature of science in the science and engineering practices and the crosscutting concepts. However, as with learning about life cycles or Newton's laws, students should be formally introduced to these concepts.

The characteristics of the nature of science can be placed into three categories: use of evidence, multiple methods of investigation, and the human factor (Crowther, Lederman, and Lederman 2009). Distribute the "The Nature of Science" handout and instruct participants to work in their groups to categorize the statements from the *NGSS* "Nature of Science Matrix."

Debrief (10 minutes): Debrief the discussion and exploration by asking participants to summarize the characteristics that are most important at their grade level by writing framing questions (e.g., How do scientists use multiple methods to collect evidence? Why is peer review important to science?). Place a labeled sheet of chart paper for each of the three categories at the front of the room and ask participants to add their framing questions to the appropriate chart.

Classroom Connections (20 minutes): The *Framework* and *NGSS* state that formal instruction is needed for students to develop an understanding of the nature of science:

The Framework *emphasizes that students must have the opportunity to stand back and reflect on how the practices contribute to the accumulation of scientific knowledge. This means, for example, that when students carry out an investigation, develop*

models, articulate questions, or engage in arguments, they should have opportunities to think about what they have done and why. They should be given opportunities to compare their own approaches to those of other students or professional scientists. Through this kind of reflection they can come to understand the importance of each practice and develop a nuanced appreciation of the nature of science. (NGSS, Appendix H, p. 7)

One way to formally introduce the nature of science to your students is through the use of anchor charts or posters. Throughout the school year, students can add examples from their science experiences to the anchor charts that are posted in the room. Participants should use the categories that they created in the previous phase of the activity to develop drafts of grade-appropriate anchor charts for their courses. Close this activity by having participants share drafts of their anchor charts with the entire group.

Next Steps

For more information on the nature of science, participants can read the NSTA position statement on the nature of science (*www.nsta.org/about/positions/natureofscience.aspx*) and the WebDigest article "Understanding the True Meaning of the Nature of Science" (Crowther, Lederman, and Lederman 2005) found at *www.nsta.org/publications/news/story.aspx?id=51055.*

ACTIVITY 10

The Nature of Science

Categorize the grade band statements in the "Nature of Science Matrix" found in *NGSS* Appendix H. These three categories are adapted from the article "Understanding the True Meaning of the Nature of Science."

Use of evidence	Multiple methods of investigations	The human factor
Scientists use evidence to explain the natural world. Explanations can change when new evidence becomes available.	*There are many ways to collect data about the natural world. Experimentation is only one of these methods.*	*Science is a human endeavor involving creativity and human subjectivity. The peer review process is a critical part of science.*
Related grade band nature of science statements from *NGSS*:	Related grade band nature of science statements from *NGSS*:	Related grade band nature of science statements from *NGSS*:

Source: Crowther, D. T., N. G. Lederman, and J. S. Lederman. 2005. Understanding the true meaning of nature of science. *Science and Children* 29 (3): 50–52.

Framing questions (e.g., How do scientists use multiple methods to collect data? Why is peer review important to science?)

Grade level: _____

7

Supporting Science Learning for All Students

I believe that in every person is a kind of circuit which resonates to intellectual discovery—and the idea is to make that resonance work.
—Carl Sagan

The achievement gap is one of the greatest challenges faced by the U.S. educational system. It is not just a classroom teacher's challenge but one that all stakeholders must address. *A Framework for K–12 Science Education* (*Framework*; NRC 2012) and the *Next Generation Science Standards* (*NGSS*; NGSS Lead States 2013) provide a giant step forward by providing accessible standards based in research on supporting the learning of all students.

The Next Generation Science Standards (NGSS) *are being developed at an historic time when major changes in education are occurring at the national level. On one hand, student demographics across the nation are changing rapidly, as teachers have seen the steady increase of student diversity in the classrooms. Yet, achievement gaps in science and other key academic indicators among demographic subgroups have persisted. On the other hand, national initiatives are emerging for a new wave of standards through the NGSS as well as* Common Core State Standards (CCSS) *for English language arts and literacy and for mathematics. As these new standards are cognitively demanding, teachers must make instructional shifts to enable all students to be college and career ready. (NGSS, Appendix D, p. 1)*

NGSS Appendix D includes seven case studies with classroom vignettes from real classrooms that highlight the teaching of distinct student groups. These groups are referred to as nondominant, not because of population size but rather due to the position in society (in this case, school) that these groups are often found. The *NGSS* writing team included classroom teachers and educational researchers with expertise in helping students from these nondominant populations succeed.

The activities in this chapter are designed to illustrate how the *NGSS* and its three dimensions of crosscutting concepts, practices of science and engineering, and disciplinary core ideas can aid all students in the science classroom.

ACTIVITY 11

Educators participate in a jigsaw activity to identify strategies for integrating the three dimensions and approaches that support the learning of students from nondominant groups.

ACTIVITY 12

Educators are introduced to three principles of Universal Design for Learning (UDL) and identify examples of modifications from the vignettes that are related to these principles. This activity is an extension of Activity 11.

ACTIVITY 13

Educators work together to complete a matrix of research-based strategies that have been found to support the learning of all students.

ACTIVITY 11

All Standards, All Students: Integrating Content, Practices, and Crosscutting Concepts

---

Approximate Length

60–90 minutes

Objectives

During this activity, participants will

- examine case studies for examples of how teachers integrate disciplinary core ideas, science and engineering practices, and crosscutting concepts; and

- identify specific strategies used in the vignettes to support student learning.

Vocabulary

- disciplinary core ideas

- science and engineering practices

- crosscutting concepts

Evidence of Learning

- Graphic organizer "Jigsaw Summary"

- Graphic organizer "Integrating the Three Dimensions"

At a Glance

This is the first in a series of three activities (Activities 11, 12, and 13) related to Appendix D, "All

Standards, All Students" from the *NGSS*. First, teachers become part of an "expert group" as they examine one case study for examples as to how the teacher in the vignette integrates disciplinary core ideas, science and engineering practices, and crosscutting concepts. Teachers also identify the specific strategies that were used to support learning in that vignette. Next, teachers form jigsaw groups (described in the Facilitator's Notes) to look for similarities across vignettes. If this activity is used in isolation, teachers can reflect individually on changes that they can make to their teaching that will aid all students in their science classrooms.

This activity is also a precursor to Activity 12, which focuses on deepening teacher understanding of Universal Design for Learning (UDL) and/or Activity 13, which helps teachers identify types of strategies that can be used to support all students' learning of science.

Facilitator's Notes

NGSS Appendix D provides a concise summary of the changing demographics of the student population in the United States and research into effective educational practices to support the learning of all students. Recognizing these changing demographics, the document uses the term "dominant" and "nondominant" to describe student diversity. Dominant populations are not always numerically superior. Instead, these populations are defined by social prestige and institutionalized privilege. Nondominant groups have traditionally been underserved by the U.S. educational system. This distinction between privileged academic background instead of purely a numerical (majority and minority) description is important because research has shown that in schools with a majority population from nondominant

groups, students from the dominant population group still see advantages from their privileged academic background. The case studies in *NGSS* Appendix D were constructed based on seven nondominant groups and illustrate instruction at a variety of grade levels. These case studies are powerful for two reasons: First, they clearly illustrate a vision for what instruction based on the *NGSS* can look like. Each case study includes a classroom vignette describing an instructional sequence that integrates disciplinary core ideas, science and engineering practices, and crosscutting concepts. The vignettes clearly show how the teacher is explicitly helping students to see how these three dimensions of the *Framework* and the *NGSS* are connected. Second, each case study describes appropriate strategies that can be used to support the learning of students from the nondominant population group featured in that case study. Research has shown that the strategies illustrated in each case study are not only effective for that nondominant population group but also support the learning of all students.

When examining these case studies, some teachers try to generalize the strategies as just "good teaching." While it is true that the classroom vignettes in these case studies do describe good teaching, encourage your workshop participants to be as specific as possible about what makes a strategy effective.

Materials

- Copies of the following handouts: "Integrating the Three Dimensions" (p. 94) and "Jigsaw Summary" (p. 95)

- Copies of (or access to) the case studies found in *NGSS* Appendix D

Procedure

Set-up: Before beginning this activity, you should be familiar with *NGSS* Appendix D and the case studies. The case studies include

- economically disadvantaged students (ninth-grade chemistry),

- students from major racial and ethnic groups (eighth-grade life science),

- students with disabilities (sixth-grade space science),

- students with limited English proficiency (second-grade Earth science),

- girls (third-grade engineering),

- students in alternative education programs (tenth- and eleventh-grade chemistry), and

- gifted and talented students (fourth-grade life science).

This activity uses a jigsaw design, so you need to determine in advance which of the case studies you will use. We have found that the activity works best with five or six case studies. Predetermine your expert (case study) and jigsaw groups. You will need one expert group for each case study. A jigsaw group is made up of one member from each of the expert groups. As an example, assume that your workshop includes 24 participants and you have decided to use six case studies. You will have six expert groups, with four members each. You will have four jigsaw groups, with six members each (one from each expert group).

Introduction (10 minutes): The *NGSS* development process included a diversity team that conducted a bias review of the standards and developed a series of case studies that include classroom vignettes describing implementation of *NGSS*. Introduce this activity by explaining to participants that they will be using a jigsaw approach to examine these case studies with two purposes in mind. First, an examination of the case studies will develop an understanding of the vision that the writing team had about what effective use of *NGSS* looks like in the classroom. Second, participants will identify specific strategies used in the case studies to support the learning of students from nondominant populations. Activate prior knowledge by having participants discuss the following prompt in their small groups: In what ways do I currently support all of my students' learning of science?

Expert Groups (20 minutes): Move the participants into expert groups and distribute the case studies and "Integrating the Three Dimensions" handout. Participants should summarize the classroom instruction exemplified by the vignettes, paying specific attention to how disciplinary core ideas, science and engineering practices, and crosscutting concepts are integrated into instruction. Participants should also describe examples of how instruction made these three dimensions explicit to students. Finally, participants should identify the specific strategies used to support the learning of students from the nondominant population group featured in the case study. As you facilitate this section of the activity, encourage participants to go beyond "it is good teaching" to include specific details of instruction.

Jigsaw Groups (30–40 minutes): After participants have summarized case studies in their expert groups, reconvene into jigsaw groups.

Each case study should be represented in the jigsaw group. Encourage participants to share their summaries of the case studies and to discuss similarities across case studies. Besides completing the "Jigsaw Summary," each jigsaw group should identify two or three big ideas that framed their discussion.

Debrief (20 minutes): Ask each jigsaw group to share the big ideas that framed the discussion in their group. After sharing, encourage each participant (or his or her small grade-level teams) to reflect on their teaching with the purpose of identifying specific opportunities for changing instruction.

Next Steps

This activity introduces participants to classroom vignettes that show *NGSS* in action. Participants also identify specific strategies that can be used to support all students' learning.

If you are interested in having participants learn more about UDL, follow this activity with Activity 12. UDL is a process that helps teachers consider the needs of their students early in the instructional planning process instead of making modifications to an instructional plan created for a generic classroom.

If you are interested in having participants learn more about the strategies identified by the *NGSS* diversity writing team that support the learning of students from nondominant populations, follow this activity with Activity 13. In Activity 13, participants will describe and generalize the strategies used in each case study.

ACTIVITY 11

Integrating the Three Dimensions

How are the three dimensions integrated and made explicit during instruction in each vignette?

Disciplinary core idea	Science and engineering practice	Crosscutting concept

Identify specific examples from the vignettes that illustrate strategies for supporting the learning of students from nondominant groups. Describe the strategy.

Example from vignette	Supporting strategy

ACTIVITY 11

Jigsaw Summary

Disciplinary core ideas	Science and engineering practices
Crosscutting concepts	Strategies to support nondominant groups

ACTIVITY 12

All Standards, All Students and Universal Design for Learning

With Stacey N. Skoning

Approximate Length

70 minutes

Objectives

During this activity, participants will

- recognize the three principles of Universal Design for Learning (UDL),

- identify instructional strategies that exemplify the three principles,

- reflect on the three principles in regard to their own classroom, and

- identify actions that could be taken to reinforce the three principles of UDL in the classrooms.

Vocabulary

- Universal Design for Learning

- reception

- expression

- engagement

Evidence of Learning

- Graphic organizer "UDL Strategy Matrix"

At a Glance

This is the second of three activities focused on understanding Appendix D, "All Standards, All Students," from the *NGSS*. This activity requires that participants have completed at least the Expert Group section of Activity 11. In this activity, participants will study the principles of UDL, identify strategies from vignettes that exemplify these principles, and reflect on actions that they can take in their own teaching.

Facilitator's Notes

Quite often, instructional planning is done with a one size fits all approach. Once the plan is complete, teachers then consider the unique needs of students in their classes and develop modifications for those students. UDL provides a framework consisting of three principles that helps teachers plan for diverse learning needs.

In principle 1 (reception), teachers consider how they can provide students with multiple representations of information. These representations should use multiple modalities (e.g., visual, auditory, kinesthetic) and include opportunities for students to interact with information in nonverbal ways. In principle 2 (expression), teachers plan multiple ways that students can take action to express what they are learning. In principle 3 (engagement), teachers plan learning activities using a variety of formats and strategies to engage and motivate students.

Materials

- Copies of (or access to) the case studies in *NGSS* Appendix D

- Copies of the following handouts: "Universal Design for Learning Essay" (pp. 99–101) and "UDL Strategies Matrix" (p. 102)

- The video, *UDL Guidelines in Practice: Grade 6 Science*, from the National Center for Universal Design for Learning (*www.udlcenter.org/resource_library/videos/udlcenter/guidelines#video4*), and a way to show it.

Procedure

Set-up: Before beginning this activity, familiarize yourself with *NGSS* Appendix D and the "Universal Design for Learning Essay" handout. Participants continue to work in jigsaw groups (from Activity 11) for most of this activity. The reflection phase of the activity should be completed either independently or within grade-level teams.

Introduction (5 minutes): Briefly introduce this activity by acknowledging the work that was done in the previous activity. Provide an overview of this activity by describing UDL as a planning approach that focuses on needs of diverse students throughout the planning process instead of waiting until the end of the process to make modifications. Explain that UDL asks teachers to think about three principles (reception, expression, engagement) related to how students interact with and communicate content. In this activity, participants will learn more about these three principles, identify examples of these three principles in the case study vignettes, and reflect on their current teaching practices.

Universal Design for Learning (25 minutes): Provide each participant with the Activity 12 handouts. Instruct participants to read the "Universal Design for Learning Essay" and summarize their understanding of each principle (reception, expression, engagement) in the first column "UDL Strategy Matrix." After participants have read the essay, show the 10-minute video, *UDL Guidelines in Practice: Grade 6 Science*. Explain to participants that they will be engaging in a similar analysis of the case study vignettes in their jigsaw groups.

UDL Strategy Matrix (20 minutes): Move participants to their jigsaw groups from Activity 11. Each jigsaw group should have an "expert" from each case study. Participants should work with their group to identify specific examples from the case study vignettes that illustrate each of the three principles. They may also include additional suggestions that could have been incorporated to make the lesson even stronger (by allowing access to an even wider range of learners).

Reflection (20 minutes): After identifying examples of each principle, participants should independently reflect on science topics they will be teaching in the near future. During reflection, each teacher should identify actions that he or she can take within each UDL principle. Encourage participants to share the strategies they would like to implement with the rest of their jigsaw group. As an alternative to individual reflections, reconvene participants in grade-level teams.

Wrap-up (Optional): Since each jigsaw group is working on the same task, you may decide it is not necessary to do a large group debriefing session. If you decide to include a large group debrief, have each jigsaw group report about the case study examples and connections to their teaching from a principle of their choice.

Next Steps

This activity introduced teachers to key concepts related to UDL. There are many resources that can be accessed to gain further knowledge about practices consistent with this approach to teaching. The following websites are good places to go for further information:

- *www.cast.org/udl*

- *www.udlcenter.org*

- *http://iris.peabody.vanderbilt.edu/module/udl*

ACTIVITY 12

Universal Design for Learning Essay

Universal design was initially developed in the field of architecture when architects discovered that it was much easier and more efficient to design buildings to be accessible from the ground up than to retrofit existing structures to be more accessible. They learned quickly that many accessibility features also made it easier for individuals who do not have disabilities to use those spaces. For example, parents with strollers use the curb-cuts to cross the street, harried travelers with their arms full use automatic door openers, and many choose to use an elevator rather than take the stairs. Everyone benefits from the inclusion of universal design into the planning process.

Just as we now consider access in the design of public spaces, all students should have access to high-quality general education curriculum. Access to such curriculum is viewed by many as a civil right (Valle and Conner 2010). The *Next Generation Science Standards* (*NGSS*; NGSS Lead States 2013) were designed to support the education of all children in science. To ensure this access, we consider three key principles of Universal Design for Learning (UDL) to plan for the many diverse learners in our classrooms. These three key principles are to consider a range of accommodations in (1) reception, (2) expression, and (3) engagement (CAST 2012; CEC 2005).

The Principle of Reception

When teaching a diverse group of students, information must be presented in more than one way. Students bring with them different understandings (and misunderstandings) of the topic. Additionally, there are many different preferences for how to access new information. The principle of reception reminds their teachers that they need to have multiple ways of presenting information so that their students are more likely to understand it. Some questions teachers should ask themselves about the principle of reception during planning include the following:

- Who are my students? What are their strengths? What do they have difficulty with?

- What specifically do they need to learn from this lesson? What is the key content/concept my students need to understand?

- What sequence should I teach the content/concept in so that understanding is more likely and so that I can scaffold the learning of my students?

- Are there multilevel or multisensory materials with which my students can engage to understand the content/concept better (Udvari-Solner, Villa, and Thousand 2005)?

- Have I considered ways to learn the information that do not rely solely on visual or auditory processes?

- Are there ways that students who are not literate can still learn the content/concept?

The Principle of Expression

Expression refers to how students demonstrate their understanding of newly learned content. Some students understand what has been taught, but have difficulty expressing that understanding when asked to do so in only one way (particularly common are difficulties with expression in written or oral language formats). It is important to have multiple methods available for students to demonstrate their understanding. Some questions teachers should ask themselves while developing lessons that address multiple means of expression include the following:

- Do I have multiple assessment measures and multilevel criteria by which I can judge student learning and progress toward meeting identified goals (Udvari-Solner, Villa, and Thousand 2005)?

- Have I included assessment measures that do not require students to write or to speak as their only means of communicating information?

- Are there practical demonstrations of learning in addition to (or in place of) more traditional quizzes or tests?

Multilevel assessment criteria: While changing assessment measures (often to those that are more performance based) to match student goals and provide a range of opportunities and ways students may express their understanding is the key to successful implementation of UDL; additional accommodations may be needed.

These accommodations may include providing adjustable levels of challenge (Peterson and Hittie 2010) or altering the criteria used to judge progress. Criteria can be differentiated by reducing the number of test items, changing the kind of test items given (true/false, multiple choice, short answer, essay, and so on) to match the way a student is able to express understanding, or adding other skills (behavioral, social, and so on) to traditional assessment routines.

The Principle of Engagement

For students to learn content, they must be engaged in the learning process throughout the lesson. All students may have the skills needed to understand the information and to share what they have learned, but if a student has not been engaged in the entire lesson, he or she will not have the same level of understanding as another student who was engaged. There are several questions that encourage increased levels of engagement that teachers should ask themselves while planning lessons. Some of these questions include the following:

- Can I use a wider range of lesson formats beyond lecture? (Lesson formats may include games, simulations, role play, demonstrations, and experiments or they may be activity-based or computer- or web-based lessons.)

- Have I varied the instructional arrangement so that students are working independently, with partners, in small groups, in cooperative groups, and in large groups to provide additional support to those who may need it?

- Are my classroom social and physical environments safe and conducive to learning and engagement?

- Have I considered specific teaching strategies to support the individual learning styles and needs of my students (Udvari-Solner, Villa, and Thousand 2005)?

Teachers who address these questions in their planning create strong lessons that are accessible to a wide range of learners. They also report spending less time planning accommodations because they are not trying to retrofit old lessons to fit new students. Their lessons are much more accommodating from the beginning.

References

Council for Exceptional Children (CEC). 2005. *Universal design for learning: A guide for teachers and education professionals*. Arlington, VA: Pearson and CEC.

Peterson, J. M., and M. M. Hittie. 2010. *Inclusive teaching: The journey toward effective schools for all learners*. 2nd ed. Boston: Pearson.

Udvari-Solner, A., R. A Villa, and J. S. Thousand. 2005. Access to the general education curriculum for all: The universal design process. In *Creating an inclusive school*, ed. R. A. Villa and J. S. Thousand, 134–155. 2nd ed. Alexandria, VA: ASCD.

Valle, J. W., and D. J. Conner. 2010. *Rethinking disability: A disability studies approach to inclusive practices*. New York: McGraw-Hill.

ACTIVITY 12

UDL Strategy Matrix

UDL principle	Specific examples from vignettes	Connections to my teaching
Reception		
Expression		
Engagement		

ACTIVITY 13

All Standards, All Students: A Strategy Matrix

With Stacey N. Skoning

Approximate Length

70 minutes

Objectives

During this activity, participants will

- understand key features of effective strategies on science learning for nondominant groups,

- identify examples of strategies from vignettes that support student science learning, and

- reflect on actions that can be taken their own classrooms to support student science learning.

Vocabulary

- nondominant groups

Evidence of Learning

- Graphic organizer "Strategy Matrix"

At a Glance

This is the last of three activities focused on understanding Appendix D, "All Standards, All Students," from the *NGSS*. This activity requires that participants have completed at least the Expert Group section of Activity 11 (p. 92). In this activity, participants will learn about the key features of effective strategies from the research literature on science learning for nondominant groups, identify examples of strategies from the vignettes that fit with these features, and reflect on actions that they can take in their teaching.

Facilitator's Notes

All of the background information needed to be successful with this activity is included in *NGSS* Appendix D. However, facilitators should be aware that some teachers gravitate toward a simplistic description of the strategies as being "just good teaching." Encourage groups to be specific in describing the characteristics of strategies included in the vignettes and why those characteristics are important for students in the nondominant group featured in that case study.

Case study 3, "Students With Disabilities and the *NGSS*," is unique when compared to the other case studies. Instead of illustrating features of effective strategies, this case study introduces two frameworks for planning instruction that is accessible to all students. Universal Design for Learning (UDL) was the focus of Activity 12. In place of these two processes, we use the following differentiation suggestions adapted from Udvari-Solner (1995):

- Changes to the instructional arrangement (grouping)

- Changes to the instructional format and style

- Changes to the classroom environment

- Changes to the materials used

Materials

- Copies of (or access to) the case studies in *NGSS* Appendix D

- Copies of (or access to) pages 6–8 of *NGSS* Appendix D (including the introduction to the Implementation of Effective Strategies and the Effective Classroom Strategies subsection)

- Copies of the handout "Strategy Matrix" (p. 105)

Procedure

Set-up: Before beginning this activity, make sure that you are familiar with *NGSS* Appendix D and the associated case studies. Participants should continue to work in jigsaw groups (from Activity 11) for most of this activity. The reflection phase of the activity should be completed either independently or within grade-level teams.

Introduction (5 minutes): Briefly introduce this activity by acknowledging the work that was done in the previous activity. Provide an overview of this activity by explaining that the *NGSS* identified a series of key features of strategies that support learning by students from nondominant populations. In this activity, participants will learn more about these key features, identify examples of these three features in the case study vignettes, and reflect on their current teaching practices.

Implementation of Effective Strategies (15 minutes): Provide each participant with Activity 12 handout, "UDL Strategy Matrix" and pages 6–8 of *NGSS* Appendix D. Participants should read the excerpt from the appendix and take notes as necessary in the first column of the handout.

Strategy Matrix (30 minutes): Move participants to their jigsaw groups from Activity 11. Each jigsaw group should have an "expert" from each case study. Participants should work with their group to identify specific examples from the case study vignettes that illustrate the key features of effective strategies for each nondominant group. For each example, participants should provide a general description of the strategy used.

Reflection (20 minutes): After identifying examples, participants should independently reflect on the science topics, they will be teaching in the near future. As they reflect on this, each teacher should identify actions that he or she can take related to the key features. Encourage participants to share the actions that they can take with the rest of their jigsaw group. As an alternative to individual reflections, reconvene participants into grade-level teams.

Wrap-up (10 minutes): Since each jigsaw group is working on the same task, you may decide that it is not necessary to do a large-group debrief. If you do decide to include a large-group debrief, have each jigsaw group report out on the case study examples and connections to their teaching from a nondominant group of their choice.

Next Steps

See Appendix 1 (p. 201) for resources related to UDL, differentiation, and supporting the learning of English language learners.

ACTIVITY 13

Strategy Matrix

Nondominant group	Specific example from a vignette	Generalized example	Opportunity
Include the defined group and the specific strategies called out in *NGSS* Appendix D.			
Economically disadvantaged students 1. Connecting to a sense of place 2. Applying students' funds of knowledge 3. Project-based learning 4. Resources and funding for science education			
Students from major racial and ethnic groups 1. Culturally relevant pedagogy 2. Community involvement and social activism 3. Multiple representation and multimodal experiences 4. School support systems, role models, and mentors			

Nondominant group	Specific example from a vignette	Generalized example	Opportunity
Students with disabilities 1. Changes to the instructional arrangement (grouping) 2. Changes to the instructional format and style 3. Changes to the classroom environment 4. Changes to the materials used			
Students with limited English proficiency 1. Literacy strategies 2. Language support strategies 3. Discourse strategies. 4. Home language support 5. Home culture connections			
Girls 1. Instructional strategies to increase achievement and perseverance 2. Curricula to support achievement, confidence, and images of successful females in science 3. Organizational (classroom and school) structure that benefits and promotes girls in science			

ACTIVITY 13

Strategy Matrix

Nondominant group	Specific example from a vignette	Generalized example	Opportunity
Students in alternative education programs. 1. Structured after-school opportunities 2. Family outreach 3. Life skills training 4. Safe learning environments 5. Individualized academic support			
Gifted and talented students. 1. Fast pacing 2. Differentiated level of challenge 3. Opportunities for self-direction 4. Strategic grouping			

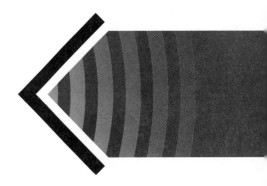

8

Fostering Discussion About Curricular Decisions

If you don't know where you are going, you might wind up someplace else.
—Yogi Berra

In today's educational world, educators face the pressure of preparing students to score at proficient or advanced levels on standardized tests, while also preparing students to be college and career ready. This can only happen with thoughtfully designed curriculum or a road map that connects standards, instruction, and assessment. This chapter addresses how to foster discussions about curricular decisions and begin to build a cohesive science road map.

The authors of the *Next Generation Science Standards* (*NGSS*; NGSS Lead States 2013) state specifically that the standards are not curriculum:

The NGSS are student outcomes and are explicitly NOT curriculum. Even though within each NGSS performance expectation Scientific and Engineering Practices (SEP) are partnered with a particular Disciplinary Core Idea (DCI) and Crosscutting Concept (CC), these intersections do not predetermine how the three are linked in the curriculum, units, lessons, or instruction; they simply clarify the expectations of what students will know and be able to do by the end of the grade or grade band. Though considering where performance expectations (PEs) will be addressed within courses is an important step in curriculum development, additional work will be needed to create coherent instructional programs that help students achieve these standards. (Appendix K, p. 5)

The activities in this chapter will introduce strategies to help educators begin to design a curriculum based on the *NGSS*.

ACTIVITY 14

Educators work in teams to identify what values and aspirations they have for science education in their district.

ACTIVITY 15

Educators work in grade-level teams to create a beginning K–5 curriculum map for their district.

ACTIVITY 16

Educators in grades 6–12 begin to identify what they currently teach that aligns with *NGSS* and what changes they may need to make in their current curriculum.

ACTIVITY 17

Educators in grades 6–12 explore possible clustering of disciplinary core ideas into course sequences.

ACTIVITY 18

Educators practice writing essential questions and big ideas using *NGSS*.

ACTIVITY 19

Educators work in teams to develop performance tasks from the performance expectations in *NGSS*.

ACTIVITY 20

Educators develop a basic unit of instruction based on *NGSS*.

ACTIVITY 14

Visioning and Values

Approximate Length

50–60 minutes

Objectives

During this activity, participants will

- identify values and aspirations for science education,

- create a list of values to drive grade/ school/district curricular decisions, and

- write an "elevator speech" to describe their values related to teaching and learning science.

Vocabulary

- elevator speech

Evidence of Learning

- List of values

- Written elevator speech

At a Glance

In this activity, teams of teachers identify what values and aspirations they have for science at their school. Participants then write an elevator speech that summarizes these values. This activity works best with teams made up of teachers (and administrators) from the same school. The activity can be completed individually, but the focus then narrows to a vision for a single course.

Facilitator's Notes

This activity is a straightforward way to help teams generate an informal vision statement to guide their work. Throughout the activity, reinforce that the purpose of identifying values is to help guide the curriculum revision process as the schools begin to align to *NGSS*. Implementation of this activity should be fast paced and lively. In addition, remind participants that they should encourage lots of ideas and refrain from judging those ideas during brainstorming sessions.

This activity uses the elevator speech method. An elevator speech is a short sales pitch that can be delivered while riding an elevator from the first to the third floor—two paragraphs, at most.

Materials

- Chart paper
- Markers
- Sentence strips (or strips of paper approximately 3″ × 12″)
- Masking tape

Procedure

Set-up: Before you begin this activity, assign participants to school teams (preferred) or grade-level groups. Begin with a stack of sentence strips (at least five per group), markers, and tape at each table. Have extra sentence strips available at a central location.

Introduction (5 minutes): Introduce the activity by explaining that teams and schools should clearly express their values before beginning any curriculum revision process. This way, curricular decisions can be made in a way that honors the values of the school community.

Newspaper Headlines (10 minutes): The purpose of the first component of this activity is to generate aspirations of what science education could look like at the participants' schools. Instruct teams to imagine that it is five years into the future and their school has successfully implemented the *NGSS*. What news headlines will local news outlets in your area write about science education at their school? For example, a headline might read, "North High School Students Showcase STEM Projects Connected to Local Industry." Teams should write these newspaper headlines on sentence strips and tape them to the wall.

Identifying Values (10 minutes): In this component of the activity, participants use the newspaper headlines to generate a list of values that they hold as a department, grade-level team, or school. These values are not constrained to just what was written in headlines. For example, one school district that we worked with placed value at the high school level on providing students with a large variety of elective choices in science. It is important to try to promote this value while making the transition to new courses aligned with *NGSS*.

Elevator Speeches (15 minutes): Finally, teams should craft an elevator speech to describe the values that they hold related to teaching and learning science. Teams should write this elevator speech on chart paper.

Gallery Walk (10 minutes): To debrief this activity, teams should either deliver their elevator speeches or be provided with time to walk around and view the speeches written by other groups.

Next Steps

Participants should refer back to the product of this activity on a regular basis during the curriculum revision process.

ACTIVITY 15

Course Mapping at the Elementary Level

Approximate Length

85 minutes

Objectives

During this activity, participants will

- explore the science topics in *NGSS* for each grade level K–5,

- identify the crosscutting concepts taught at each grade level K–5,

- compare current science topics taught with science topics in *NGSS* for their grade level, and

- identify curricular changes for grades K–5.

Vocabulary

- crosscutting concepts

- science topics

Evidence of Learning

- Graphic organizer "Topics and Crosscutting Concepts"

- Graphic organizer "Brainstorming Topics and Crosscutting Concepts"

- "Curriculum Mapping Feedback" form

- "*NGSS* Mapping, Grades K–5" chart

At a Glance

The *NGSS* are organized by grade at the elementary level. The *NGSS* have been thoughtfully scaffolded so that standards are developmentally appropriate for students to build knowledge over their elementary educational years. This activity allows participants to explore that scaffolding and see where topics fit within grades K–5. Participants will gain a better understanding of content that is being taught at the various grade levels. It should be noted, that this activity is a big picture perspective meant to allow participants to gain a better understanding of what is taught at each elementary grade.

Facilitator's Notes

As mentioned at the beginning of Chapter 8 (p. 109), the writers of *NGSS* are clear that the *NGSS* are outcomes, not a curriculum.

The NGSS *are student outcomes and are explicitly NOT curriculum. Even though within each* NGSS *performance expectation Scientific and Engineering Practices (SEP) are partnered with a particular Disciplinary Core Idea (DCI) and Crosscutting Concept (CC), these intersections do not predetermine how the three are linked in the curriculum, units, lessons, or instruction; they simply clarify the expectations of what students will know and be able to do by the end of the grade or grade band. Though considering where performance expectations (PEs) will be addressed within courses is an important step in curriculum development, additional work will be needed to create coherent instructional programs that help students achieve these standards. (NGSS, Appendix K, p. 5)*

Because the *NGSS* state performance expectations for each grade, it is important that educators understand the scaffolding and see the progression of what students will know and be able to do by the end of each grade. This activity is an exploratory activity that allows participants to begin understanding the scaffolding through exploring the topics and crosscutting concepts at each grade level.

Materials

- A copy of the *NGSS* for participants' grade level

- Copies of the following handouts: "Topics and Crosscutting Concepts" (p. 115), "Brainstorming Topics and Crosscutting Concepts" (p. 116), and "Curriculum Mapping Feedback" (p. 117)

- Large topical map available in the room to write on "*NGSS* Mapping, Grades K–5" (example at end of this activity)

- Markers

Procedure

Set-up: Participants should be in grade-level groups prior to beginning the activity. A large copy of the "*NGSS* Mapping Grades K–5" handout should be placed on wall in the room for participants to write on. This activity works best if participants are placed in groups of three to four educators at the same grade level. This activity also requires a minimum of one group per grade level, K–5.

Introduction (5 minutes): Explain that participants will be examining the *NGSS* to determine topics to be instructed at each grade level. Grade-level teams will determine the topics and

crosscutting concepts that have been aligned with their grade.

Group Work (30 minutes): Participants should explore the *NGSS* and complete the "Topics and Crosscutting Concepts" handout. All groups should be instructed to add their information to the "*NGSS* Mapping, Grades K–5" topical map on the wall.

Discussion (10 minutes): As a whole group, discuss implications for the topics and crosscutting concepts at each of group's grade levels. Are there surprises? Are there areas for concern? Do participants see the scaffolding?

Group Work (30 minutes): Participants should return to original groups to begin brainstorming how *NGSS* supports what they are already doing in their classrooms and changes that they will have to make. They should also explore how the crosscutting concept supports other disciplines. For example, if they are instructing patterns in science, where else might patterns fit in and lead to a greater understanding of patterns for students. The purpose of this exploratory time is for participants to make connections to what they already do in the classroom and realize that they do not have to throw out everything that they currently instruct to implement *NGSS* successfully. Participants may want to use the graphic organizer "Brainstorming Topics and Crosscutting Concepts" to organize their thoughts and discussion.

Debrief (10 minutes): Close this activity by allowing grade-level groups to share some of their thoughts on moving forward with *NGSS*. Have participants complete the "Curriculum Mapping Feedback" handout, and collect them.

Next Steps

Review the filled-out "Curriculum Mapping Feedback" handouts. This document will give direction to your professional development planning. Some teachers have difficulty letting go of curriculum that they previously taught or are intimidated by new topics. This feedback form should help alert you to these needs. Participants need to move beyond "this is what I do" to "this is what should be done" to help students to be successful in their school.

ACTIVITY 15

Topics and Crosscutting Concepts

Grade	Domain	Topics	Crosscutting concepts
	Life science		
	Physical science		
	Earth and space sciences		
	Engineering, technology, and applications of science		

ACTIVITY 15

Brainstorming Topics and Crosscutting Concepts

Grade Level _____

Topics	Already do	New or need to add

Crosscutting concept	Other disciplines (units) to integrate crosscutting concepts

ACTIVITY 15
Curriculum Mapping
Feedback

Aha—what was something that you learned?	Concerns?
Topics that you will keep at your grade level	New topics for your grade level

We will need help with ...

Grade Level _____

ACTIVITY 15

NGSS Mapping, Grades K–5

District _____

Grade	Domain	Topics	Crosscutting concepts
K	L.S.		
	P.S.		
	E.S./S.S.		
	E.T.A.		
1	L.S.		
	P.S.		
	E.S./S.S		
	E.T.A.		
2	L.S.		
	P.S.		
	E.S./S.S		
	E.T.A.		
3	L.S.		
	P.S.		
	E.S./S.S		
	E.T.A.		
4	L.S.		
	P.S.		
	E.S./S.S		
	E.T.A.		
5	L.S.		
	P.S.		
	E.S./S.S		
	E.T.A.		

E.S./S.S. = Earth and Space Sciences; E.T.A. = Engineering, Technology, and Application of Science; L.S. = Life Science; P.S. = Physical Science

NATIONAL SCIENCE TEACHERS ASSOCIATION

ACTIVITY 16

Plus, Minus, Delta (Grades 6–12)

Approximate Length

45 minutes

Objectives

During this activity, participants will

- explore *NGSS* to align what they currently teach with *NGSS*, and

- identify science curricular changes.

Vocabulary

- DCI progression matrix

Evidence of Learning

- Graphic organizer "What Do We Teach?"

At a Glance

In this activity, participants examine the *NGSS* standards for their grade band (6–8, 9–12) specific to the topics that they are currently teaching. Participants determine which current topics are well aligned with *NGSS*, which might be possible to deemphasize or cut, and what changes may be necessary. The purpose of the activity is to help teachers of science begin to think about possible changes to content and topics within courses.

Facilitator's Notes

One goal of *A Framework for K–12 Science Education* (*Framework*) and *NGSS* was to identify a smaller set of coherent ideas that students can explore in depth. The content in *NGSS* represents the big ideas that form the foundation for understanding science—content that all students should understand by the time they leave high school. As a result, most courses will need to be revised to shrink the number of topics that are covered each year. This is always a challenging effort because some of the topics that are eliminated are topics that teachers feel passionate about or find enjoyable to teach. However, it is important that we unburden our curriculum so that we can realize the vision of the *Framework* and *NGSS*.

It is important to remember that this activity begins the discussion. However, teachers should not make large changes to their curriculum (e.g., adding or eliminating topics) without permission.

Materials

- "What Do We Teach?" handout (p. 121)

- Access to *NGSS*. As an alternative to providing participants access to the entire *NGSS* document, you can provide them with the "DCI Progression Matrix" (*NGSS*, Appendix E)

Procedure

Set-up: Before you begin this activity, assign participants to grade-level or content-alike (preferred) teams. Pass out the "What Do We Teach?" handout.

Introduction (5 minutes): Explain that groups will explore how the content they currently teach is aligned with *NGSS*. Reinforce that this is the beginning of the process to identify possible curricular changes but should not be acted upon without school or district approval. Teachers are responsible for the district-adopted curriculum.

Group Work (25 minutes): Provide time for participants to work within their groups. They should identify areas of agreement (+) between standards documents and areas that are not well aligned (–). Finally, participants should identify the major changes that they see themselves making as they align their courses with *NGSS*.

Debrief (15 minutes): Provide time for participants to summarize and share their work.

Next Steps

This activity serves to begin a discussion about curricular changes. The course mapping activities delve into curricular divisions with more depth. For more information on unburdening the curriculum, check out Chapter 7, "Unburdening the Curriculum," in *Designs for Science Literacy* (*www. project2061.org/publications/designs/ch7intro.htm*).

ACTIVITY 16

What Do We Teach?

+: What are you doing that is already well aligned with NGSS?	**–**: What are you doing that could be minimized or eliminated?

△: What changes to your course topics might be necessary?

ACTIVITY 17

Course Mapping for Middle and High School

Approximate Length

60 minutes

Objectives

During this activity, participants will

- explore disciplinary core ideas from *NGSS*, and

- cluster disciplinary core ideas to begin building courses.

Vocabulary

- course mapping

Evidence of Learning

- "Disciplinary Core Ideas in the *Next Generation Science Standards* (*NGSS*) Final Release" chart

- "Course Mapping Feedback" form

At a Glance

This activity provides a structure for beginning to think through the process of clustering performance expectations into a course sequence at the middle and high school level. Since the science and engineering practices and crosscutting concepts need to be reinforced every year, we focus this planning on using the disciplinary core idea matrix designed by NSTA.

Facilitator's Notes

The *NGSS* are written as a set of performance expectations that describe what all students should know and be able to do by the time they leave high school. Performance expectations are guidelines for assessment, not a curriculum or a mandate for a specific set of courses. The *NGSS* provide organizational structures based on disciplinary core ideas or topics, which package related performance expectations together into bundles. This bundling reinforces the importance of viewing performance expectations as related statements instead of viewing them in isolation.

Because the standards at the middle school and high school levels are not delineated into specific courses, schools and districts will need to complete a course mapping process. Throughout the process, teachers need to keep in mind that the *NGSS* are intended for all students—by the time students graduate from high school, they should all be able to competently perform all of the performance expectations. These standards provide a strong foundation for students to enter a college or career as scientifically literate citizens. However, it is also important to realize that many students will take advanced coursework that includes concepts that are more sophisticated than what is in *NGSS*. *NGSS* is not meant to replace or constrain advanced elective courses.

To begin developing course maps, it is important to review *NGSS* Appendix K, "Model Course Building in Middle and High School." This appendix identifies three different course structures. The domain model includes one course focusing on physical science, another on life science, and a third on Earth and space sciences. The modified domain model is geared toward a traditional sequence of biology, chemistry, and physics, and integrates Earth and space sciences performance expectations

Caution

This activity uses only the disciplinary core ideas to map courses. Remind participants multiple times during this activity that courses will eventually be built from performance expectations that integrate the selected disciplinary core ideas with science and engineering practices and crosscutting concepts. Science and engineering practices and connections to crosscutting concepts need to be developed throughout the year during each course.

into those three courses. A third course sequence provides an integrated approach. This conceptual model sequences core ideas so that they build on each other throughout middle and high school but does not bind them by specific discipline.

Appendix K only provides three possibilities. However, there are many other ways that a school or district could sequence core ideas into courses. It might be easy to keep the "status quo" and grab onto the domain model because that is what the school is currently doing. That may turn out to be the best option for a school, but now is the chance to really think through the types of experiences that students could have. Challenge participants to identify multiple ways to design courses before they settle on an approach. Consider what the school community values, and think outside of the proverbial box. For example, if a school values providing students with a variety of electives, the administrators should attempt to find a model that maintains student choice.

During the course mapping process, it is important to be creative and collaborative and to give adequate time to develop multiple options. For example, could you create a course focused on remote sensing? Humans use devices to extend the senses. A course centered around this concept could introduce many space science topics, core ideas related to waves and light, properties of materials, nuclear science, and even some biology concepts.

A second example could be to develop a structure that provides students with choices and electives. Performance expectations could be bundled into different categories. Then, within each category, the performance expectations can be met in different ways. In a category focused on ecosystems and human impacts, you might have one course on lake ecosystems and a second on forest ecosystems. Students would then be required to take one course from each category.

Materials

- Chart paper

- Markers

- Clear tape

- Disciplinary core idea cards: To prepare these cards, cut out the 6–8 and 9–12 portion of the disciplinary core ideas matrix "Disciplinary Core Ideas in the *Next Generation Science Standards* (*NGSS*) Final Release" (pp. 125–138). Each card should contain one disciplinary core idea. Each group should have a full set of cards for their grade band. If you plan on doing this activity multiple times, you may find it helpful to tape each disciplinary core idea to a small index card or duplicate on card stock prior to cutting apart. We have also found it helpful to color code the cards by domain. For example, Earth and space sciences as blue, life science as yellow, and physical science as green. This allows participants to easily see overlaps and ways to integrate the areas.

- "Course Mapping Feedback" handout (p. 139)

Procedure

Set-up: Before beginning this activity, determine grade band groups for participants and prepare one set of cards for each grade band. This activity works best when participants work in grade band groups (i.e., 6–8 and 9–12). You should also be familiar with *NGSS* Appendix K, "Model Course Mapping in Middle and High School."

Introduction (15 minutes): The purpose of this activity is to begin thinking about how the disciplinary core ideas from *NGSS* can be sequenced into courses at the middle and high school level. Participants will use the disciplinary core idea cards to create a variety of courses. Use the Facilitator's Notes section above to describe the course maps provided in *NGSS* Appendix K (also see model slides 1–3). Encourage groups to be creative and to think outside of traditional course structures. For example, slide 4 shows a series of five semester-long courses that include a course that integrates Earth and life sciences and a course that consists of physical science (chemistry) and life science topics.

Group Work (30 minutes): Participants work in groups to cluster disciplinary core ideas into courses at the middle and high school level. At the end of this work, groups should summarize their work (either by taping cards or writing) on chart paper. These chart paper summaries should be taped to a wall or otherwise made visible to the whole group.

Gallery Walk (10 minutes): Provide time for participants to view the work done by other groups. Encourage them to take notes on ideas of interest and to leave comments and suggestions on the summaries.

Debrief (10 minutes): Ask participants to reflect on the activity. Talk about next steps to complete course mapping. Have participants complete the "Course Mapping Feedback" handout. Collect this handout to inform future curriculum work.

Next Steps

Participants can use the *NGSS* database (*www. nextgenscience.org/search-performance-expectations*) to generate more detail for the potential course sequences that they develop. The database allows you to select individual performance expectations and then automatically creates the foundation and connection boxes.

Disciplinary Core Ideas in the Next Generation Science Standards (NGSS) Final Release

Topic	Primary School (Grades K-2)	Elementary School (Grades 3-5)	Middle School (Grades 6-8)	High School (Grades 9-12)
Life Science				
LS1: From Molecules to Organisms: Structures and Processes				
LS1.A: Structure and Function	• All organisms have external parts. Different animals use their body parts in different ways to see, hear, grasp objects, protect themselves, move from place to place, and seek, find, and take in food, water and air. Plants also have different parts (roots, stems, leaves, flowers, fruits) that help them survive and grow. (1-LS1-1)	• Plants and animals have both internal and external structures that serve various functions in growth, survival, behavior, and reproduction. (4-LS1-1)	• All living things are made up of cells, which is the smallest unit that can be said to be alive. An organism may consist of one single cell (unicellular) or many different numbers and types of cells (multicellular). (MS-LS1-1) • Organisms reproduce, either sexually or asexually, and transfer their genetic information to their offspring. (secondary to MSLS3-2) • Within cells, special structures are responsible for particular functions, and the cell membrane forms the boundary that controls what enters and leaves the cell. (MS-LS1-2) • In multicellular organisms, the body is a system of multiple interacting subsystems. These subsystems are groups of cells that work together to form tissues and organs that are specialized for particular body functions. (MS-LS1-3)	• Systems of specialized cells within organisms help them perform the essential functions of life. (HS-LS1-1) • All cells contain genetic information in the form of DNA molecules. Genes are regions in the DNA that contain the instructions that code for the formation of proteins, which carry out most of the work of cells. (HS-LS1-1) (secondary to HS-LS3-1) • Multicellular organisms have a hierarchical structural organization, in which any one system is made up of numerous parts and is itself a component of the next level. (HS-LS1-2) • Feedback mechanisms maintain a living system's internal conditions within certain limits and mediate behaviors, allowing it to remain alive and functional even as external conditions change within some range. Feedback mechanisms can encourage (through positive feedback) or discourage (negative feedback) what is going on inside the living system. (HS-LS1-3)
LS1.B: Growth and Development of Organisms	• Adult plants and animals can have young. In many kinds of animals, parents and the offspring themselves engage in behaviors that help the offspring to survive. (1-LS1-2)	• Reproduction is essential to the continued existence of every kind of organism. Plants and animals have unique and diverse life cycles. (3-LS1-1)	• Animals engage in characteristic behaviors that increase the odds of reproduction. (MS-LS1-4) • Plants reproduce in a variety of ways, sometimes depending on animal behavior and specialized features for reproduction. (MS-LS1-4) • Genetic factors as well as local conditions affect the growth of the adult plant. (MS-LS1-5)	• In multicellular organisms individual cells grow and then divide via a process called mitosis, thereby allowing the organism to grow. The organism begins as a single cell (fertilized egg) that divides successively to produce many cells, with each parent cell passing identical genetic material (two variants of each chromosome pair) to both daughter cells. Cellular division and differentiation produce and maintain a complex organism, composed of systems of tissues and organs that work together to meet the needs of the whole organism. (HS-LS1-4)
LS1.C: Organization for Matter and Energy Flow in Organisms	• All animals need food in order to live and grow. They obtain their food from plants or from other animals. Plants need water and light to live and grow. (K-LS1-1)	• Food provides animals with the materials they need for body repair and growth and the energy they need to maintain body warmth and for motion. (secondary to 5-PS3-1) • Plants acquire their material for growth chiefly from air and water. (5-LS1-1)	• Plants, algae (including phytoplankton), and many microorganisms use the energy from light to make sugars (food) from carbon dioxide from the atmosphere and water through the process of photosynthesis, which also releases oxygen. These sugars can be used immediately or stored for growth or later use. (MS-LS1-6) • Within individual organisms, food moves through a series of chemical reactions in which it is broken down and rearranged to form new molecules, to support growth, or to release energy. (MS-LS1-7)	• The process of photosynthesis converts light energy to stored chemical energy by converting carbon dioxide plus water into sugars plus released oxygen. (HS-LS1-5) • The sugar molecules thus formed contain carbon, hydrogen, and oxygen: their hydrocarbon backbones are used to make amino acids and other carbon-based molecules that can be assembled into larger molecules (such as proteins or DNA), used for example to form new cells. (HS-LS1-6) • As matter and energy flow through different organizational levels of living systems, chemical elements are recombined in different ways to form different products. (HS-LS1-6),(HS-LS1-7) • As a result of these chemical reactions, energy is transferred from one system of interacting molecules to another and release energy to the surrounding environment and to maintain body temperature. Cellular respiration is a chemical process whereby the bonds of food molecules and oxygen molecules are broken and new compounds are formed that can transport energy to muscles. (HS-LS1-7)
LS1.D: Information Processing	• Animals have body parts that capture and convey different kinds of information needed for growth and survival. Animals respond to these inputs with behaviors that help them survive. Plants also respond to some external inputs. (1-LS1-1)	• Different sense receptors are specialized for particular kinds of information, which may be then processed by the animal's brain. Animals are able to use their perceptions and memories to guide their actions. (4-LS1-2)	• Each sense receptor responds to different inputs (electromagnetic, mechanical, chemical), transmitting them as signals that travel along nerve cells to the brain. The signals are then processed in the brain, resulting in immediate behaviors or memories. (MS-LS1-8)	

Matrix Developed by NSTA 5/9/2013

1

LS2: Ecosystems: Interactions, Energy, and Dynamics

Topic	Primary School (Grades K-2)	Elementary School (Grades 3-5)	Middle School (Grades 6-8)	High School (Grades 9-12)
LS2.A: Interdependent Relationships in Ecosystems	• Plants depend on water and light to grow. (2-LS2-1) • Plants depend on animals for pollination or to move their seeds around. (2-LS2-2)	• The food of almost any kind of animal can be traced back to plants. Organisms are related in food webs in which some animals eat plants for food and other animals eat the animals that eat plants. Some organisms, such as fungi and bacteria, break down dead organisms (both plants or plants parts and animals) and therefore operate as "decomposers." Decomposition eventually restores (recycles) some materials back to the soil. Organisms can survive only in environments in which their particular needs are met. A healthy ecosystem is one in which multiple species of different types are each able to meet their needs in a relatively stable web of life. Newly introduced species can damage the balance of an ecosystem. (5-LS2-1)	• Organisms, and populations of organisms, are dependent on their environmental interactions both with other living things and with nonliving factors. (MS-LS2-1) • In any ecosystem, organisms and populations with similar requirements for food, water, oxygen, or other resources may compete with each other for limited resources, access to which consequently constrains their growth and reproduction. (MS-LS2-1) • Growth of organisms and population increases are limited by access to resources. (MS-LS2-1) • Similarly, predatory interactions may reduce the number of organisms or eliminate whole populations of organisms. Mutually beneficial interactions, in contrast, may become so interdependent that each organism requires the other for survival. Although the species involved in these competitive, predatory, and mutually beneficial interactions vary across ecosystems, the patterns of interactions of organisms with their environments, both living and nonliving, are shared. (MS-LS2-2)	• Ecosystems have carrying capacities, which are limits to the numbers of organisms and populations they can support. These limits result from such factors as the availability of living and nonliving resources and from such challenges such as predation, competition, and disease. Organisms would have the capacity to produce populations of great size were it not for the fact that environments and resources are finite. This fundamental tension affects the abundance (number of individuals) of species in any given ecosystem. (HS-LS2-1),(HSLS2-2)
LS2.B: Cycles of Matter and Energy Transfer in Ecosystems		• Matter cycles between the air and soil and among plants, animals, and microbes as these organisms live and die. Organisms obtain gases, and water, from the environment, and release waste matter (gas, liquid, or solid) back into the environment. (5-LS2-1)	• Food webs are models that demonstrate how matter and energy is transferred between producers, consumers, and decomposers as the three groups interact within an ecosystem. Transfers of matter into and out of the physical environment occur at every level. Decomposers recycle nutrients from dead plant or animal matter back to the soil in terrestrial environments or to the water in aquatic environments. The atoms that make up the organisms in an ecosystem are cycled repeatedly between the living and nonliving parts of the ecosystem. (MS-LS2-3)	• Photosynthesis and cellular respiration (including anaerobic processes) provide most of the energy for life processes. (HS-LS2-3) • Plants or algae form the lowest level of the food web. At each link upward in a food web, only a small fraction of the matter consumed at the lower level is transferred upward, to produce growth and release energy in cellular respiration at the higher level. Given this inefficiency, there are generally fewer organisms at higher levels of a food web. Some matter reacts to release energy for life functions, some matter is stored in newly made structures, and much is discarded. The chemical elements that make up the molecules of organisms pass through food webs and into and out of the atmosphere and soil, and they are combined and recombined in different ways. At each link in an ecosystem, matter and energy are conserved. (HS-LS2-4) • Photosynthesis and cellular respiration are important components of the carbon cycle, in which carbon is exchanged among the biosphere, atmosphere, oceans, and geosphere through chemical, physical, geological, and biological processes. (HS-LS2-5)
LS2.C: Ecosystem Dynamics, Functioning, and Resilience		• When the environment changes in ways that affect a place's physical characteristics, temperature, or availability of resources, some organisms survive and reproduce, others move to new locations, yet others move into the transformed environment, and some die. (secondary to 3-LS4-4)	• Ecosystems are dynamic in nature; their characteristics can vary over time. Disruptions to any physical or biological component of an ecosystem can lead to shifts in all its populations. (MS-LS2-4) • Biodiversity describes the variety of species found in Earth's terrestrial and oceanic ecosystems. The completeness or integrity of an ecosystem's biodiversity is often used as a measure of its health. (MS-LS2-5)	• A complex set of interactions within an ecosystem can keep its numbers and types of organisms relatively constant over long periods of time under stable conditions. If a modest biological or physical disturbance to an ecosystem occurs, it may return to its more or less original status (i.e., the ecosystem is resilient), as opposed to becoming a very different ecosystem. Extreme fluctuations in conditions or the size of any population, however, can challenge the functioning of ecosystems in terms of resources and habitat availability. (HS-LS2-2),(HS-LS2-6) • Moreover, anthropogenic changes (induced by human activity) in the environment—including habitat destruction, pollution, introduction of invasive species, overexploitation, and climate change—can disrupt an ecosystem and threaten the survival of some species. (HS-LS2-7)

Based on the Disciplinary Core Ideas in the *NGSS* Final Release (May 2013)

Topic	Primary School (Grades K-2)	Elementary School (Grades 3-5)	Middle School (Grades 6-8)	High School (Grades 9-12)
LS2.D: Social Interactions and Group Behavior		Being part of a group helps animals obtain food, defend themselves, and cope with changes. Groups may serve different functions and vary dramatically in size (Note: Moved from K-2). (3-LS2-1)	Changes in biodiversity can influence humans' resources, such as food, energy, and medicines, as well as ecosystem services that humans rely on—for example, water purification and recycling. (secondary to MS-LS2-5)	Group behavior has evolved because membership can increase the chances of survival for individuals and their genetic relatives. (HSLS2-8)

LS3: Heredity: Inheritance and Variation of Traits

Topic	Primary School (Grades K-2)	Elementary School (Grades 3-5)	Middle School (Grades 6-8)	High School (Grades 9-12)
LS3.A: Inheritance of Traits	Young animals are very much, but not exactly, like their parents. Plants also are very much, but not exactly, like their parents. (1-LS3-1)	• Many characteristics of organisms are inherited from their parents. (3-LS3-1) • Other characteristics result from individuals' interactions with the environment, which can range from diet to learning. Many characteristics involve both inheritance and environment. (3-LS3-2)	• Genes are located in the chromosomes of cells, with each chromosome pair containing two variants of each of many distinct genes. Each distinct gene chiefly controls the production of specific proteins, which in turn affects the traits of the individual. Changes (mutations) to genes can result in changes to proteins, which can affect the structures and functions of the organism and thereby change traits. (MS-LS3-1) • Variations of inherited traits between parent and offspring arise from genetic differences that result from the subset of chromosomes (and therefore genes) inherited. (MS-LS3-2)	• Each chromosome consists of a single very long DNA molecule, and each gene on the chromosome is a particular segment of that DNA. The instructions for forming species' characteristics are carried in DNA. All cells in an organism have the same genetic content, but the genes used (expressed) by the cell may be regulated in different ways. Not all DNA codes for a protein; some segments of DNA are involved in regulatory or structural functions, and some have no as-yet known function. (HS-LS3-1)
LS3.B: Variation of Traits	Individuals of the same kind of plant or animal are recognizable as similar but can also vary in many ways. (1-LS3-1)	• Different organisms vary in how they look and function because they have different inherited information. (3-LS3-1) • The environment also affects the traits that an organism develops. (3-LS3-2)	• In sexually reproducing organisms, each parent contributes half of the genes acquired (at random) by the offspring. Individuals have two of each chromosome and hence two alleles of each gene, one acquired from each parent. These versions may be identical or may differ from each other. (MS-LS3-2) • In addition to variations that arise from sexual reproduction, genetic information can be altered because of mutations. Though rare, mutations may result in changes to the structure and function of proteins. Some changes are beneficial, others harmful, and some neutral to the organism. (MS-LS3-1)	• In sexual reproduction, chromosomes can sometimes swap sections during the process of meiosis (cell division), thereby creating new genetic combinations and thus more genetic variation. Although DNA replication is tightly regulated and remarkably accurate, errors do occur and result in mutations, which are also a source of genetic variation. Environmental factors can also cause mutations in genes, and viable mutations are inherited. (HS-LS3-2) • Environmental factors also affect expression of traits, and hence affect the probability of occurrences of traits in a population. Thus the variation and distribution of traits observed depends on both genetic and environmental factors. (HS-LS3-2),(HS-LS3-3)

LS4: Biological Evolution: Unity and Diversity

Topic	Primary School (Grades K-2)	Elementary School (Grades 3-5)	Middle School (Grades 6-8)	High School (Grades 9-12)
LS4.A: Evidence of Common Ancestry and Diversity		• Some kinds of plants and animals that once lived on Earth are no longer found anywhere. (Note: moved from K-2) (3-LS4-1) • Fossils provide evidence about the types of organisms that lived long ago and also about the nature of their environments. (3-LS4-1)	• The collection of fossils and their placement in chronological order (e.g., through the location of the sedimentary layers in which they are found or through radioactive dating) is known as the fossil record. It documents the existence, diversity, extinction, and change of many life forms throughout the history of life on Earth. (MS-LS4-1) • Anatomical similarities and differences between various organisms living today and between them and organisms in the fossil record, enable the reconstruction of evolutionary history and the inference of lines of evolutionary descent. (MS-LS4-2) • Comparison of the embryological development of different species also reveals similarities that show relationships not evident in the fully-formed anatomy. (MS-LS4-3)	• Genetic information provides evidence of evolution. DNA sequences vary among species, but there are many overlaps; in fact, the ongoing branching that produces multiple lines of descent can be inferred by comparing the DNA sequences of different organisms. Such information is also derivable from the similarities and differences in amino acid sequences and from anatomical and embryological evidence. (HS-LS4-1)
LS4.B: Natural Selection		Sometimes the differences in characteristics between individuals of the same species provide advantages in surviving, finding mates, and reproducing. (3-LS4-2)	• Natural selection leads to the predominance of certain traits in a population, and the suppression of others. (MS-LS4-4) • In artificial selection, humans have the capacity to influence certain characteristics of organisms by selective breeding. One can choose desired parental traits determined by genes, which are then passed on to offspring. (MS-LS4-5)	• Natural selection occurs only if there is both (1) variation in the genetic information between organisms in a population and (2) variation in the expression of that genetic information—that is, trait variation—that leads to differences in performance among individuals. (HS-LS4-2),(HS-LS4-3) • The traits that positively affect survival are more likely to be reproduced, and thus are more common in the population. (HS-LS4-3)

Matrix Developed by NSTA 5/9/2013

3

Topic	Primary School (Grades K-2)	Elementary School (Grades 3-5)	Middle School (Grades 6-8)	High School (Grades 9-12)
LS4.C: Adaptation		• For any particular environment, some kinds of organisms survive well, some survive less well, and some cannot survive at all. (3-LS4-3)	• Adaptation by natural selection acting over generations is one important process by which species change over time in response to changes in environmental conditions. Traits that support successful survival and reproduction in the new environment become more common; those that do not become less common. Thus, the distribution of traits in a population changes. (MS-LS4-6)	• Evolution is a consequence of the interaction of four factors: (1) the potential for a species to increase in number, (2) the genetic variation of individuals in a species due to mutation and sexual reproduction, (3) competition for an environment's limited supply of the resources that individuals need in order to survive and reproduce, and (4) the ensuing proliferation of those organisms that are better able to survive and reproduce in that environment. (HS-LS4-2) • Natural selection leads to adaptation, that is, to a population dominated by organisms that are anatomically, behaviorally, and physiologically well suited to survive and reproduce in a specific environment. That is, the differential survival and reproduction of organisms in a population that have an advantageous heritable trait leads to an increase in the proportion of individuals in future generations that have the trait and to a decrease in the proportion of individuals that do not. (HS-LS4-3),(HS-LS4-4) • Adaptation also means that the distribution of traits in a population can change when conditions change. (HS-LS4-3) • Changes in the physical environment, whether naturally occurring or human induced, have thus contributed to the expansion of some species, the emergence of new distinct species as populations diverge under different conditions, and the decline—and sometimes the extinction—of some species. (HS-LS4-5),(HS-LS4-6) • Species become extinct because they can no longer survive and reproduce in their altered environment. If members cannot adjust to change that is too fast or drastic, the opportunity for the species' evolution is lost. (HS-LS4-5)
LS4.D: Biodiversity and Humans	• There are many different kinds of living things in any area, and they exist in different places on land and in water. (2-LS4-1)	• Populations live in a variety of habitats, and change in those habitats affects the organisms living there. (3-LS4-4)		• Biodiversity is increased by the formation of new species (speciation) and decreased by the loss of species (extinction). (secondary to HSLS2-7) • Humans depend on the living world for the resources and other benefits provided by biodiversity. But human activity is also having adverse impacts on biodiversity through overpopulation, overexploitation, habitat destruction, pollution, introduction of invasive species, and climate change. Thus sustaining biodiversity so that ecosystem functioning and productivity are maintained is essential to supporting and enhancing life on Earth. Sustaining biodiversity also aids humanity by preserving landscapes of recreational or inspirational value. (secondary to HS-LS2-7) (HS-LS4-6)

Based on the Disciplinary Core Ideas in the *NGSS* Final Release (May 2013)

4

Earth and Space Science

ESS1: Earth's Place in the Universe

Topic	Primary School (Grades K-2)	Elementary School (Grades 3-5)	Middle School (Grades 6-8)	High School (Grades 9-12)
ESS1.A: The Universe and Its Stars	• Patterns of the motion of the sun, moon, and stars in the sky can be observed, described, and predicted. (1-ESS1-1)	• The sun is a star that appears larger and brighter than other stars because it is closer. Stars range greatly in their distance from Earth. (5-ESS1-1)	• Patterns of the apparent motion of the sun, the moon, and stars in the sky can be observed, described, predicted, and explained with models. (MS-ESS1-1) • Earth and its solar system are part of the Milky Way galaxy, which is one of many galaxies in the universe. (MS-ESS1-2)	• The star called the sun is changing and will burn out over a lifespan of approximately 10 billion years. (HS-ESS1-1) • The study of stars' light spectra and brightness is used to identify compositional elements of stars, their movements, and their distances from Earth. (HS-ESS1-2),(HS-ESS1-3) • The Big Bang theory is supported by observations of distant galaxies receding from our own, of the measured composition of stars and non-stellar gases, and of the maps of spectra of the primordial radiation (cosmic microwave background) that still fills the universe. (HSESS1-2) • Other than the hydrogen and helium formed at the time of the Big Bang, nuclear fusion within stars produces all atomic nuclei lighter than and including iron, and the process releases electromagnetic energy. Heavier elements are produced when certain massive stars achieve a supernova stage and explode. (HS-ESS1-2),(HS-ESS1-3)
ESS1.B: Earth and the Solar System	• Seasonal patterns of sunrise and sunset can be observed, described, and predicted. (1-ESS1-2)	• The orbits of Earth around the sun and of the moon around Earth, together with the rotation of Earth about an axis between its North and South poles, cause observable patterns. These include day and night; daily changes in the length and direction of shadows; and different positions of the sun, moon, and stars at different times of the day, month, and year. (5-ESS1-2)	• The solar system consists of the sun and a collection of objects, including planets, their moons, and asteroids that are held in orbit around the sun by its gravitational pull on them. (MS-ESS1-2),(MSESS1-3) • This model of the solar system can explain eclipses of the sun and the moon. Earth's spin axis is fixed in direction over the short-term but tilted relative to its orbit around the sun. The seasons are a result of that tilt and are caused by the differential intensity of sunlight falling on different areas of Earth across the year. (MS-ESS1-1) • The solar system appears to have formed from a disk of dust and gas, drawn together by gravity. (MS-ESS1-2)	• Kepler's laws describe common features of the motions of orbiting objects, including their elliptical paths around the sun. Orbits may change due to the gravitational effects from, or collisions with, other objects in the solar system. (HS-ESS1-4) • Cyclical changes in the shape of Earth's orbit around the sun, together with changes in the tilt of the planet's axis of rotation, both occurring over hundreds of thousands of years, have altered the intensity and distribution of sunlight falling on the earth. These phenomena cause a cycle of ice ages and other gradual climate changes. (secondary to HS-ESS2-4)
ESS1.C: The History of Planet Earth	• Some events happen very quickly; others occur very slowly, over a time period much longer than one can observe. (2-ESS1-1)	• Local, regional, and global patterns of rock formations reveal changes over time due to earth forces, such as earthquakes. The presence and location of certain fossil types indicate the order in which rock layers were formed. (4-ESS1-1)	• The geologic time scale interpreted from rock strata provides a way to organize Earth's history. Analyses of rock strata and the fossil record provide only relative dates, not an absolute scale. (MS-ESS1-4) • Tectonic processes continually generate new ocean sea floor at ridges and destroy old sea floor at trenches. (HS.ESS1.C GBE) (secondary to MS-ESS2-3)	• Continental rocks, which can be older than 4 billion years, are generally much older than the rocks of the ocean floor, which are less than 200 million years old. (HS-ESS1-5) • Although active geologic processes, such as plate tectonics and erosion, have destroyed or altered most of the very early rock record on Earth, other objects in the solar system, such as lunar rocks, asteroids, and meteorites, have changed little over billions of years. Studying these objects can provide information about Earth's formation and early history. (HS-ESS1-6)

INTRODUCING TEACHERS + ADMINISTRATORS TO THE *NGSS*
A PROFESSIONAL DEVELOPMENT FACILITATOR'S GUIDE

Topic	Primary School (Grades K-2)	Elementary School (Grades 3-5)	Middle School (Grades 6-8)	High School (Grades 9-12)
ESS2: Earth's Systems				
ESS2.A: Earth Materials and Systems	• Wind and water can change the shape of the land. (2-ESS2-1)	• Rainfall helps to shape the land and affects the types of living things found in a region. Water, ice, wind, living organisms, and gravity break rocks, soils, and sediments into smaller particles and move them around. (4-ESS2-1) • Earth's major systems are the geosphere (solid and molten rock, soil, and sediments), the hydrosphere (water and ice), the atmosphere (air), and the biosphere (living things, including humans). These systems interact in multiple ways to affect Earth's surface materials and processes. The ocean supports a variety of ecosystems and organisms, shapes landforms, and influences climate. Winds and clouds in the atmosphere interact with the landforms to determine patterns of weather. (5-ESS2-1)	• All Earth processes are the result of energy flowing and matter cycling within and among the planet's systems. This energy is derived from the sun and Earth's hot interior. The energy that flows and matter that cycles produce chemical and physical changes in Earth's materials and living organisms. (MS-ESS2-1) • The planet's systems interact over scales that range from microscopic to global in size, and they operate over fractions of a second to billions of years. These interactions have shaped Earth's history and will determine its future. (MS-ESS2-2)	• Earth's systems, being dynamic and interacting, cause feedback effects that can increase or decrease the original changes. (HSESS2-1),(HS-ESS2-2) • Evidence from deep probes and seismic waves, reconstructions of historical changes in Earth's surface and its magnetic field, and an understanding of physical and chemical processes lead to a model of Earth with a hot but solid inner core, a liquid outer core, a solid mantle and crust. Motions of the mantle and its plates occur primarily through thermal convection, which involves the cycling of matter due to the outward flow of energy from Earth's interior and gravitational movement of denser materials toward the interior. (HS-ESS2-3) • The geological record shows that changes to global and regional climate can be caused by interactions among changes in the sun's energy output or Earth's orbit, tectonic events, ocean circulation, volcanic activity, glaciers, vegetation, and human activities. These changes can occur on a variety of time scales from sudden (e.g., volcanic ash clouds) to intermediate (ice ages) to very long-term tectonic cycles. (HS-ESS2-4)
ESS2.B: Plate Tectonics and Large-Scale System Interactions	• Maps show where things are located. One can map the shapes and kinds of land and water in any area. (2-ESS2-2)	• The locations of mountain ranges, deep ocean trenches, ocean floor structures, earthquakes, and volcanoes occur in patterns. Most earthquakes and volcanoes occur in bands that are often along the boundaries between continents and oceans. Major mountain chains form inside continents or near their edges. Maps can help locate the different land and water features areas of Earth. (4-ESS2-2)	• Maps of ancient land and water patterns, based on investigations of rocks and fossils, make clear how Earth's plates have moved great distances, collided, and spread apart. (MS-ESS2-3)	• The radioactive decay of unstable isotopes continually generates new energy within Earth's crust and mantle, providing the primary source of the heat that drives mantle convection. Plate tectonics can be viewed as the surface expression of mantle convection. (HS-ESS2-3) • Plate tectonics is the unifying theory that explains the past and current movements of the rocks at Earth's surface and provides a framework for understanding its geologic history. (ESS2.B Grade 8 GBE) (HS-ESS2-1) (secondary to HS-ESS1-5)
ESS2.C: The Roles of Water in Earth's Surface Processes	• Water is found in the ocean, rivers, lakes, and ponds. Water exists as solid ice and in liquid form. (2-ESS2-3)	• Nearly all of Earth's available water is in the ocean. Most fresh water is in glaciers or underground; only a tiny fraction is in streams, lakes, wetlands, and the atmosphere. (5-ESS2-2)	• Water continually cycles among land, ocean, and atmosphere via transpiration, evaporation, condensation and crystallization, and precipitation, as well as downhill flows on land. (MS-ESS2-4) • The complex patterns of the changes and the movement of water in the atmosphere, determined by winds, landforms, and ocean temperatures and currents, are major determinants of local weather patterns. (MSESS2-5) • Global movements of water and its changes in form are propelled by sunlight and gravity. (MS-ESS2-4) • Variations in density due to variations in temperature and salinity drive a global pattern of interconnected ocean currents. (MS-ESS2-6) • Water's movements—both on the land and underground—cause weathering and erosion, which change the land's surface features and create underground formations. (MS-ESS2-2)	• The abundance of liquid water on Earth's surface and its unique combination of physical and chemical properties are central to the planet's dynamics. These properties include water's exceptional capacity to absorb, store, and release large amounts of energy, transmit sunlight, expand upon freezing, dissolve and transport materials, and lower the viscosities and melting points of rocks. (HS-ESS2-5)

Based on the Disciplinary Core Ideas in the *NGSS* Final Release (May 2013)

NATIONAL SCIENCE TEACHERS ASSOCIATION

Topic	Primary School (Grades K-2)	Elementary School (Grades 3-5)	Middle School (Grades 6-8)	High School (Grades 9-12)
ESS2.D: Weather and Climate	• Weather is the combination of sunlight, wind, snow or rain, and temperature in a particular region at a particular time. People measure these conditions to describe and record the weather and to notice patterns over time. (K-ESS2-1)	• Scientists record patterns of the weather across different times and areas so that they can make predictions about what kind of weather might happen next. (3-ESS2-1) • Climate describes a range of an area's typical weather conditions and the extent to which those conditions vary over years. (3-ESS2-2)	• Weather and climate are influenced by interactions involving sunlight, the ocean, the atmosphere, ice, landforms, and living things. These interactions vary with latitude, altitude, and local and regional geography, all of which can affect oceanic and atmospheric flow patterns. (MS-ESS2-6) • Because these patterns are so complex, weather can only be predicted probabilistically. (MS-ESS2-5) • The ocean exerts a major influence on weather and climate by absorbing energy from the sun, releasing it over time, and globally redistributing it through ocean currents. (MS-ESS2-6)	• The foundation for Earth's global climate systems is the electromagnetic radiation from the sun, as well as its reflection, absorption, storage, and redistribution among the atmosphere, ocean, and land systems, and this energy's re-radiation into space. (HS-ESS2-4) • Gradual atmospheric changes were due to plants and other organisms that captured carbon dioxide and released oxygen. (HS-ESS2-6),(HS-ESS2-7) • Changes in the atmosphere due to human activity have increased carbon dioxide concentrations and thus affect climate. (HS-ESS2-6),(HS-ESS2-4) • Current models predict that, although future regional climate changes will be complex and varied, average global temperatures will continue to rise. The outcomes predicted by global climate models strongly depend on the amounts of human-generated greenhouse gases added to the atmosphere each year and by the ways in which these gases are absorbed by the ocean and biosphere. (secondary to HSESS3-6)
ESS2.E: Biogeology	• Plants and animals can change their environment. (KESS2-2)	• Living things affect the physical characteristics of their regions. (4-ESS2-1)		• The many dynamic and delicate feedbacks between the biosphere and other Earth systems cause a continual co-evolution of Earth's surface and the life that exists on it. (HS-ESS2-7)

Matrix Developed by NSTA 5/9/2013

Topic	Primary School (Grades K-2)	Elementary School (Grades 3-5)	Middle School (Grades 6-8)	High School (Grades 9-12)
ESS3: Earth and Human Activity				
ESS3.A: Natural Resources	• Living things need water, air, and resources from the land, and they live in places that have the things they need. Humans use natural resources for everything they do. (K-ESS3-1)	• Energy and fuels that humans use are derived from natural sources, and their use affects the environment in multiple ways. Some resources are renewable over time, and others are not. (4-ESS3-1)	• Humans depend on Earth's land, ocean, atmosphere, and biosphere for many different resources. Minerals, fresh water, and biosphere resources are limited, and many are not renewable or replaceable over human lifetimes. These resources are distributed unevenly around the planet as a result of past geologic processes. (MS-ESS3-1)	• Resource availability has guided the development of human society. (HS-ESS3-1) • All forms of energy production and other resource extraction have associated economic, social, environmental, and geopolitical costs and risks as well as benefits. New technologies and social regulations can change the balance of these factors. (HS-ESS3-2)
ESS3.B: Natural Hazards	• Some kinds of severe weather are more likely than others in a given region. Weather scientists forecast severe weather so that the communities can prepare for and respond to these events. (K-ESS3-2)	• A variety of natural hazards result from natural processes. Humans cannot eliminate natural hazards but can take steps to reduce their impacts. (3-ESS3-1) (4-ESS3-2.)	• Mapping the history of natural hazards in a region, combined with an understanding of related geologic forces can help forecast the locations and likelihoods of future events. (MS-ESS3-2)	• Natural hazards and other geologic events have shaped the course of human history; [they] have significantly altered the sizes of human populations and have driven human migrations. (HS-ESS3-1)
ESS3.C: Human Impacts on Earth Systems	• Things that people do to live comfortably can affect the world around them. But they can make choices that reduce their impacts on the land, water, air, and other living things. (K-ESS3-3) (secondary to K-ESS2-2)	• Human activities in agriculture, industry, and everyday life have had major effects on the land, vegetation, streams, ocean, air, and even outer space. But individuals and communities are doing things to help protect Earth's resources and environments. (5-ESS3-1)	• Human activities have significantly altered the biosphere, sometimes damaging or destroying natural habitats and causing the extinction of other species. But changes to Earth's environments can have different impacts (negative and positive) for different living things. (MS-ESS3-3) • Typically as human populations and per-capita consumption of natural resources increase, so do the negative impacts on Earth unless the activities and technologies involved are engineered otherwise. (MSESS3-3),(MS-ESS3-4)	• The sustainability of human societies and the biodiversity that supports them requires responsible management of natural resources. (HS-ESS3-3) • Scientists and engineers can make major contributions by developing technologies that produce less pollution and waste and that preclude ecosystem degradation. (HS-ESS3-4)
ESS3.D: Global Climate Change		•	• Human activities, such as the release of greenhouse gases from burning fossil fuels, are major factors in the current rise in Earth's mean surface temperature (global warming). Reducing the level of climate change and reducing human vulnerability to whatever climate changes do occur depend on the understanding of climate science, engineering capabilities, and other kinds of knowledge, such as understanding of human behavior and on applying that knowledge wisely in decisions and activities. (MS-ESS3-5)	• Though the magnitudes of human impacts are greater than they have ever been, so too are human abilities to model, predict, and manage current and future impacts. (HS-ESS3-5) • Through computer simulations and other studies, important discoveries are still being made about how the ocean, the atmosphere, and the biosphere interact and are modified in response to human activities. (HS-ESS3-6)

Physical Science

PS1: Matter and Its Interactions

Topic	Primary School (Grades K-2)	Elementary School (Grades 3-5)	Middle School (Grades 6-8)	High School (Grades 9-12)
PS1.A: Structure and Properties of Matter	• Different kinds of matter exist and many of them can be either solid or liquid, depending on temperature. Matter can be described and classified by its observable properties. (2-PS1-1) • Different properties are suited to different purposes. (2-PS1-2),(2-PS1-3) • A great variety of objects can be built up from a small set of pieces. (2-PS1-3)	• Matter of any type can be subdivided into particles that are too small to see, but even then the matter still exists and can be detected by other means. A model shows that gases are made from matter particles that are too small to see and are moving freely around in space can explain many observations, including the inflation and shape of a balloon; the effects of air on larger particles or objects. (5-PS1-1) • The amount (weight) of matter is conserved when it changes form, even in transitions in which it seems to vanish. (5-PS1-2) • Measurements of a variety of properties can be used to identify materials. (Boundary: At this grade level, mass and weight are not distinguished, and no attempt is made to define the unseen particles or explain the atomic-scale mechanism of evaporation and condensation.) (5-PS1-3)	• Substances are made from different types of atoms, which combine with one another in various ways. Atoms form molecules that range in size from two to thousands of atoms. (MS-PS1-1) • Each pure substance has characteristic physical and chemical properties (for any bulk quantity under given conditions) that can be used to identify it. (MS-PS1-2), (MS-PS1-3) • Gases and liquids are made of molecules or inert atoms that are moving about relative to each other. (MS-PS1-4) • In a liquid, the molecules are constantly in contact with others; in a gas, they are widely spaced except when they happen to collide. In a solid, atoms are closely spaced and may vibrate in position but do not change relative locations. (MS-PS1-4) • Solids may be formed from molecules, or they may be extended structures with repeating subunits (e.g., crystals). (MS-PS1-1) • The changes of state that occur with variations in temperature or pressure can be described and predicted using these models of matter. (MS-PS1-4)	• Each atom has a charged substructure consisting of a nucleus, which is made of protons and neutrons, surrounded by electrons. (HS-PS1-1) • The periodic table orders elements horizontally by the number of protons in the atom's nucleus and places those with similar chemical properties in columns. The repeating patterns of this table reflect patterns of outer electron states. (HS-PS1-1),(HS-PS1-2) • The structure and interactions of matter at the bulk scale are determined by electrical forces within and between atoms. (HS-PS1-3),(secondary to HS-PS2-6) • Stable forms of matter are those in which the electric and magnetic field energy is minimized. A stable molecule has less energy than the same set of atoms separated; one must provide at least this energy in order to take the molecule apart. (HS-PS1-4)
PS1.B: Chemical Reactions	• Heating or cooling a substance may cause changes that can be observed. Sometimes these changes are reversible, and sometimes they are not. (2-PS1-4)	• When two or more different substances are mixed, a new substance with different properties may be formed. (5-PS1-4) • No matter what reaction or change in properties occurs, the total weight of the substances does not change. (Boundary: Mass and weight are not distinguished at this grade level.) (5-PS1-2)	• Substances react chemically in characteristic ways. In a chemical process, the atoms that make up the original substances are regrouped into different molecules, and these new substances have different properties from those of the reactants. (MS-PS1-2),(MS-PS1-3),(MS-PS1-5) • The total number of each type of atom is conserved, and thus the mass does not change. (MS-PS1-5) • Some chemical reactions release energy, others store energy. (MS-PS1-6)	• Chemical processes, their rates, and whether or not energy is stored or released can be understood in terms of the collisions of molecules and the rearrangements of atoms into new molecules, with consequent changes in the sum of all bond energies in the set of molecules that are matched by changes in kinetic energy. (HSPS1-4),(HS-PS1-5) • In many situations, a dynamic and condition-dependent balance between a reaction and the reverse reaction determines the numbers of all types of molecules present. (HS-PS1-6) • The fact that atoms are conserved, together with knowledge of the chemical properties of the elements involved, can be used to describe and predict chemical reactions. (HS-PS1-2),(HS-PS1-7)
PS1.C: Nuclear Processes				• Nuclear processes, including fusion, fission, and radioactive decays of unstable nuclei, involve release or absorption of energy. The total number of neutrons plus protons does not change in any nuclear process. (HSPS1-8) • Spontaneous radioactive decays follow a characteristic exponential decay law. Nuclear lifetimes allow radiometric dating to be used to determine the ages of rocks and other materials. (secondary to HS-ESS1-5),(secondary to HS-ESS1-6)

Matrix Developed by NSTA 5/9/2013

Topic	Primary School (Grades K-2)	Elementary School (Grades 3-5)	Middle School (Grades 6-8)	High School (Grades 9-12)
PS2: Motion and Stability: Forces and Interactions				
PS2.A: Forces and Motion	• Pushes and pulls can have different strengths and directions. (KPS2-1),(K-PS2-2) • Pushing or pulling on an object can change the speed or direction of its motion and can start or stop it. (K-PS2-1),(K-PS2-2)	• Each force acts on one particular object and has both strength and a direction. An object at rest typically has multiple forces acting on it, but they add to give zero net force on the object. Forces that do not sum to zero can cause changes in the object's speed or direction of motion. (Boundary: Qualitative and conceptual, but not quantitative addition of forces are used at this level.) (3-PS2-1) • The patterns of an object's motion in various situations can be observed and measured; when that past motion exhibits a regular pattern, future motion can be predicted from it. (Boundary: Technical terms, such as magnitude, velocity, momentum, and vector quantity, are not introduced at this level, but the concept that some quantities need both size and direction to be described is developed.) (3-PS2-2)	• For any pair of interacting objects, the force exerted by the first object on the second object is equal in strength to the force that the second object exerts on the first, but in the opposite direction (Newton's third law). (MS-PS2-1) • The motion of an object is determined by the sum of the forces acting on it; if the total force on the object is not zero, its motion will change. The greater the mass of the object, the greater the force needed to achieve the same change in motion. For any given object, a larger force causes a larger change in motion. (MS-PS2-2) • All positions of objects and the directions of forces and motions must be described in an arbitrarily chosen reference frame and arbitrarily chosen units of size. In order to share information with other people, these choices must also be shared. (MSPS2-2)	• Newton's second law accurately predicts changes in the motion of macroscopic objects. (HS-PS2-1) • Momentum is defined for a particular frame of reference; it is the mass times the velocity of the object. In any system, total momentum is always conserved. (HS-PS2-2) • If a system interacts with objects outside itself, the total momentum of the system can change; however, any such change is balanced by changes in the momentum of objects outside the system. (HS-PS2-2),(HS-PS2-3)
PS2.B: Types of Interactions	• When objects touch or collide, they push on one another and can change motion. (K-PS2-1)	• Objects in contact exert forces on each other. (3-PS2-1) • Electric, and magnetic forces between a pair of objects do not require that the objects be in contact. The sizes of the forces in each situation depend on the properties of the objects and their distances apart and, for forces between two magnets, on their orientation relative to each other. (3-PS2-3),(3-PS2-4) • The gravitational force of Earth acting on an object near Earth's surface pulls that object toward the planet's center. (5-PS2-1)	• Electric and magnetic (electromagnetic) forces can be attractive or repulsive, and their sizes depend on the magnitudes of the charges, currents, or magnetic strengths involved and on the distances between the interacting objects. (MS-PS2-3) • Gravitational forces are always attractive. There is a gravitational force between any two masses, but it is very small except when one or both of the objects have large mass—e.g., Earth and the sun. (MS-PS2-4) • Forces that act at a distance (electric and magnetic) can be explained by fields that extend through space and can be mapped by their effect on a test object (a ball, a charged object, or a magnet, respectively). (MS-PS2-5)	• Newton's law of universal gravitation and Coulomb's law provide the mathematical models to describe and predict the effects of gravitational and electrostatic forces between distant objects. (HS-PS2-4) • Forces at a distance are explained by fields (gravitational, electric, and magnetic) permeating space that can transfer energy through space. Magnets or electric currents cause magnetic fields; electric charges or changing magnetic fields cause electric fields. (HS-PS2-4),(HS-PS2-5) • Attraction and repulsion between electric charges at the atomic scale explain the structure, properties, and transformations of matter, as well as the contact forces between material objects. (HS-PS2-6),(secondary to HS-PS1-1),(secondary to HS-PS1-3)
PS2.C: Stability and Instability in Physical Systems				…and "electrical energy" may mean energy stored in a battery or energy transmitted by electric currents. (secondary to HS-PS2-5)

PS3: Energy

Topic	Primary School (Grades K-2)	Elementary School (Grades 3-5)	Middle School (Grades 6-8)	High School (Grades 9-12)
PS3.A: Definitions of Energy		• The faster a given object is moving, the more energy it possesses. (4-PS3-1) • Energy can be moved from place to place by moving objects or through sound, light, or electric currents. (4-PS3-2),(4-PS3-3)	• Motion energy is properly called kinetic energy; it is proportional to the mass of the moving object and grows with the square of its speed. (MS-PS3-1) • A system of objects may also contain stored (potential) energy, depending on their relative positions. (MS-PS3-2) • Temperature is a measure of the average kinetic energy of particles of matter. The relationship between the temperature and the total energy of a system depends on the types, states, and amounts of matter present. (MS-PS3-3),(MS-PS3-4) • The term "heat" as used in everyday language refers both to thermal motion (the motion of atoms or molecules within a substance) and radiation (particularly infrared and light). In science, heat is used only for this second meaning; it refers to energy transferred when two objects or systems are at different temperatures. (secondary to MS-PS1-4) • Temperature is not a measure of energy; the relationship between the temperature and the total energy of a system depends on the types, states, and amounts of matter present. (secondary to MS-PS1-4)	• Energy is a quantitative property of a system that depends on the motion and interactions of matter and radiation within that system. That there is a single quantity called energy is due to the fact that a system's total energy is conserved, even as, within the system, energy is continually transferred from one object to another and between its various possible forms. (HSPS3-1),(HS-PS3-2) • At the macroscopic scale, energy manifests itself in multiple ways, such as in motion, sound, light, and thermal energy. (HSPS3-2) (HS-PS3-3) • These relationships are better understood at the microscopic scale, at which all of the different manifestations of energy can be modeled as either motions of particles or energy stored in fields (which mediate interactions between particles). This last concept includes radiation, a phenomenon in which energy stored in fields moves across space. (HS-PS3-2)
PS3.B: Conservation of Energy and Energy Transfer	• Sunlight warms Earth's surface. (K-PS3-1),(K-PS3-2)	• Energy is present whenever there are moving objects, sound, light, or heat. When objects collide, energy can be transferred from one object to another, thereby changing their motion. In such collisions, some energy is typically also transferred to the surrounding air; as a result, the air gets heated and sound is produced. (4-PS3-2),(4-PS3-3) • Light also transfers energy from place to place. (4-PS3-2) • Energy can also be transferred from place to place by electric currents, which can then be used locally to produce motion, sound, heat, or light. The currents may have been produced to begin with by transforming the energy of motion into electrical energy. (4-PS3-2),(4-PS3-4)	• When the motion energy of an object changes, there is inevitably some other change in energy at the same time. (MS-PS3-5) • The amount of energy transfer needed to change the temperature of a matter sample by a given amount depends on the nature of the matter, the size of the sample, and the environment. (MS-PS3-4) • Energy is spontaneously transferred out of hotter regions or objects and into colder ones. (MS-PS3-3)	• Conservation of energy means that the total change of energy in any system is always equal to the total energy transferred into or out of the system. (HS-PS3-1) • Energy cannot be created or destroyed, but it can be transported from one place to another and transferred between systems. (HS-PS3-1),(HS-PS3-4) • Mathematical expressions, which quantify how the stored energy in a system depends on its configuration (e.g. relative positions of charged particles, compression of a spring) and how kinetic energy depends on mass and speed, allow the concept of conservation of energy to be used to predict and describe system behavior. (HS-PS3-1) • The availability of energy limits what can occur in any system. (HS-PS3-1) • Uncontrolled systems always evolve toward more stable states—that is, toward more uniform energy distribution (e.g., water flows downhill, objects hotter than their surrounding environment cool down). (HS-PS3-4)
PS3.C: Relationship Between Energy and Forces	• A bigger push or pull makes things go faster. (secondary to K-PS2-1)	• When objects collide, the contact forces transfer energy so as to change the objects' motions. (4-PS3-3)	• When two objects interact, each one exerts a force on the other that can cause energy to be transferred to or from the object. (MS-PS3-2)	• When two objects interacting through a field change relative position, the energy stored in the field is changed. (HS-PS3-5)

INTRODUCING TEACHERS + ADMINISTRATORS TO THE *NGSS*
A PROFESSIONAL DEVELOPMENT FACILITATOR'S GUIDE

Matrix Developed by NSTA 5/9/2013

Topic	Primary School (Grades K-2)	Elementary School (Grades 3-5)	Middle School (Grades 6-8)	High School (Grades 9-12)
PS3.D: Energy in Chemical Processes and Everyday Life		• The expression "produce energy" typically refers to the conversion of stored energy into a desired form for practical use. (4-PS3-4) • The energy released [from] food was once energy from the sun that was captured by plants in the chemical process that forms plant matter (from air and water). (5-PS3-1)	• The chemical reaction by which plants produce complex food molecules (sugars) requires an energy input (i.e., from sunlight) to occur. In this reaction, carbon dioxide and water combine to form carbon-based organic molecules and release oxygen. (secondary to MS-LS1-6) • Cellular respiration in plants and animals involve chemical reactions with oxygen that release stored energy. In these processes, complex molecules containing carbon react with oxygen to produce carbon dioxide and other materials. (secondary to MS-LS1-7)	• Although energy cannot be destroyed, it can be converted to less useful forms—for example, to thermal energy in the surrounding environment. (HS-PS3-3),(HS-PS3-4) • Solar cells are human-made devices that likewise capture the sun's energy and produce electrical energy. (secondary to HS-PS4-5) • The main way that solar energy is captured and stored on Earth is through the complex chemical process known as photosynthesis. (secondary to HS-LS2-5) • Nuclear Fusion processes in the center of the sun release the energy that ultimately reaches Earth as radiation. (secondary to HS-ESS1-1)

Based on the Disciplinary Core Ideas in the *NGSS* Final Release (May 2013)

NATIONAL SCIENCE TEACHERS ASSOCIATION

Topic	Primary School (Grades K-2)	Elementary School (Grades 3-5)	Middle School (Grades 6-8)	High School (Grades 9-12)
PS4: Waves and Their Applications in Technologies for Information Transfer				
PS4.A: Wave Properties	• Sound can make matter vibrate, and vibrating matter can make sound. (1-PS4-1)	• Waves, which are regular patterns of motion, can be made in water by disturbing the surface. When waves move across the surface of deep water, the water goes up and down in place; it does not move in the direction of the wave except when the water meets the beach. (Note: This grade band endpoint was moved from K–2.) (4-PS4-1) • Waves of the same type can differ in amplitude (height of the wave) and wavelength (spacing between wave peaks). (4-PS4-1)	• A simple wave has a repeating pattern with a specific wavelength, frequency, and amplitude. (MS-PS4-1) • A sound wave needs a medium through which it is transmitted. (MS-PS4-2)	• The wavelength and frequency of a wave are related to one another by the speed of travel of the wave, which depends on the type of wave and the medium through which it is passing. (HS-PS4-1) • Information can be digitized (e.g., a picture stored as the values of an array of pixels); in this form, it can be stored reliably in computer memory and sent over long distances as a series of wave pulses. (HS-PS4-2),(HSPS4-5) • [From the 3–5 grade band endpoints] Waves can add or cancel one another as they cross, depending on their relative phase (i.e., relative position of peaks and troughs of the waves), but they emerge unaffected by each other. (Boundary: The discussion at this grade level is qualitative only; it can be based on the fact that two different sounds can pass a location in different directions without getting mixed up.) (HS-PS4-3) • Geologists use seismic waves and their reflection at interfaces between layers to probe structures deep in the planet. (secondary to HS-ESS2-3)
PS4.B: Electromagnetic Radiation	• Objects can be seen only when light is available to illuminate them. Some objects give off their own light. (1-PS4-2) • Some materials allow light to pass through them, others allow only some light through and others block all the light and create a dark shadow on any surface beyond them, where the light cannot reach. Mirrors can be used to redirect a light beam. (Boundary: The idea that light travels from place to place is developed through experiences with light sources, mirrors, and shadows, but no attempt is made to discuss the speed of light.) (1-PS4-3)	• An object can be seen when light reflected from its surface enters the eyes. (4-PS4-2)	• When light shines on an object, it is reflected, absorbed, or transmitted through the object, depending on the object's material and the frequency (color) of the light. (MS-PS4-2) • The path that light travels can be traced as straight lines, except at surfaces between different transparent materials (e.g., air and water, air and glass) where the light path bends. (MS-PS4-2) • A wave model of light is useful for explaining brightness, color, and the frequency-dependent bending of light at a surface between media. (MS-PS4-2) • However, because light can travel through space, it cannot be a matter wave, like sound or water waves. (MS-PS4-2)	• Electromagnetic radiation (e.g., radio, microwaves, light) can be modeled as a wave of changing electric and magnetic fields or as particles called photons. The wave model is useful for explaining many features of electromagnetic radiation, and the particle model explains other features. (HS-PS4-3) • When light or longer wavelength electromagnetic radiation is absorbed in matter, it is generally converted into thermal energy (heat). Shorter wavelength electromagnetic radiation (ultraviolet, X-rays, gamma rays) can ionize atoms and cause damage to living cells. (HS-PS4-4) • Photovoltaic materials emit electrons when they absorb light of a high-enough frequency. (HS-PS4-5) • Atoms of each element emit and absorb characteristic frequencies of light. These characteristics allow identification of the presence of an element, even in microscopic quantities. (secondary to HS-ESS1-2)
PS4.C: Information Technologies and Instrumentation	• People also use a variety of devices to communicate (send and receive information) over long distances. (1-PS4-4)	• Digitized information transmitted over long distances without significant degradation. High-tech devices, such as computers or cell phones, can receive and decode information—convert it from digitized form to voice—and vice versa. (4-PS4-3)	• Digitized signals (sent as wave pulses) are a more reliable way to encode and transmit information. (MS-PS4-3)	• Multiple technologies based on the understanding of waves and their interactions with matter are part of everyday experiences in the modern world (e.g., medical imaging, communications, scanners) and in scientific research. They are essential tools for producing, transmitting, and capturing signals and for storing and interpreting the information contained in them. (HS-PS4-5)

Matrix Developed by NSTA 5/9/2013

Engineering, Technology, and the Application of Science

ETS1: Engineering Design

Topic	Primary School (Grades K-2)	Elementary School (Grades 3-5)	Middle School (Grades 6-8)	High School (Grades 9-12)
ETS1.A: Defining and Delimiting an Engineering Problem	• A situation that people want to change or create can be approached as a problem to be solved through engineering. Such problems may have many acceptable solutions. (K-2-ETS1-1) (secondary to K-PS2-2) • Asking questions, making observations, and gathering information are helpful in thinking about problems. (K-2-ETS1-1) (secondary to K-ESS3-2) • Before beginning to design a solution, it is important to clearly understand the problem. (K-2-ETS1-1)	• Possible solutions to a problem are limited by available materials and resources (constraints). The success of a designed solution is determined by considering the desired features of a solution (criteria). Different proposals for solutions can be compared on the basis of how well each one meets the specified criteria for success or how well each takes the constraints into account. (3-5-ETS1-1) (secondary to 4-PS3-4)	• The more precisely a design task's criteria and constraints can be defined, the more likely it is that the designed solution will be successful. Specification of constraints includes consideration of scientific principles and other relevant knowledge that is likely to limit possible solutions. (MS-ETS1-1) (secondary to MS-PS3-3)	• Criteria and constraints also include satisfying any requirements set by society, such as taking issues of risk mitigation into account, and they should be quantified to the extent possible and stated in such a way that one can tell if a given design meets them. (HS-ETS1-1) (secondary to HS-PS2-3) (secondary to HS-PS3-3) • Humanity faces major global challenges today, such as the need for supplies of clean water and food or for energy sources that minimize pollution, which can be addressed through engineering. These global challenges also may have manifestations in local communities. (HS-ETS1-1)
ETS1.B: Developing Possible Solutions	• Designs can be conveyed through sketches, drawings, or physical models. These representations are useful in communicating ideas for a problem's solutions to other people. (K-2-ETS1-1) (secondary to K-ESS3-3) (secondary to 2-LS2-2)	• Research on a problem should be carried out before beginning to design a solution. Testing a solution involves investigating how well it performs under a range of likely conditions. (3-5-ETS1-2) • At whatever stage, communicating with peers about proposed solutions is an important part of the design process, and shared ideas can lead to improved designs. (3-5-ETS1-2) • Tests are often designed to identify failure points or difficulties, which suggest the elements of the design that need to be improved. (3-5-ETS1-3) • Testing a solution involves investigating how well it performs under a range of likely conditions. (secondary to 4-ESS3-2)	• A solution needs to be tested, and then modified on the basis of the test results, in order to improve it. (MS-ETS1-4) (secondary to MS-PS1-6) • There are systematic processes for evaluating solutions with respect to how well they meet criteria and constraints of a problem. (MS-ETS1-2), (MS-ETS1-3) (secondary to MS-PS3-3) (secondary to MS-LS2-5) • Sometimes parts of different solutions can be combined to create a solution that is better than any of its predecessors. (MS-ETS1-3) • Models of all kinds are important for testing solutions. (MS-ETS1-4)	• When evaluating solutions it is important to take into account a range of constraints including cost, safety, reliability and aesthetics and to consider social, cultural and environmental impacts. (secondary to HS-LS2-7) (secondary to HS-LS4-6) (secondary to HS-ESS3-2),(secondary HS-ESS3-4) (HS-ETS1-3) • Both physical models and computers can be used in various ways to aid in the engineering design process. Computers are useful for a variety of purposes, such as running simulations to test different ways of solving a problem or to see which one is most efficient or economical; and in making a persuasive presentation to a client about how a given design will meet his or her needs. (HS-ETS1-4) (secondary to HS-LS4-6)
ETS1.C: Optimizing the Design Solution	• Because there is always more than one possible solution to a problem, it is useful to compare and test designs. (K-2-ETS1-1) (secondary to 2-ESS2-1)	• Different solutions need to be tested in order to determine which of them best solves the problem, given the criteria and the constraints. (3-5-ETS1-3) (secondary to 4-PS4-3)	• Although one design may not perform the best across all tests, identifying the characteristics of the design that performed the best in each test can provide useful information for the redesign process—that is, some of the characteristics may be incorporated into the new design. (MS-ETS1-3) (secondary to MS-PS1-6) • The iterative process of testing the most promising solutions and modifying what is proposed on the basis of the test results leads to greater refinement and ultimately to an optimal solution. (MSETS1-4) (secondary to MS-PS1-6)	• Criteria may need to be broken down into simpler ones that can be approached systematically, and decisions about the priority of certain criteria over others (tradeoffs) may be needed. (HSETS1-2) (secondary to HS-PS1-6) (secondary to HS-PS2-3)

NATIONAL SCIENCE TEACHERS ASSOCIATION

Based on the Disciplinary Core Ideas in the *NGSS Final Release* (May 2013)

14

ACTIVITY 17
Course Mapping Feedback

Aha—what was something that you learned?	Concerns?
Topics that you will keep at your grade level	New topics for your grade level

We will need help with ...

Grade Level (s) _____ or Course _____

ACTIVITY 18

Essential Questions and Crosscutting Concepts

Approximate Length

45–50 minutes

Objectives

During this activity, participants will

- learn what big ideas and essential questions are;

- explore the relationship between disciplinary core ideas, crosscutting concepts, essential questions, and big ideas; and

- write essential questions using crosscutting concepts.

Vocabulary

- big ideas

- essential questions

Evidence of Learning

- Graphic organizer "What Do We Want to Learn? Essential Questions"

At a Glance

This activity is designed to demonstrate one process for designing essential questions or "what do we want to learn" using the *NGSS* crosscutting concepts. Participants will be involved in a short discussion on "why we need to do this" and "how we do it." Next, participants will try and write several essential questions on their own to demonstrate their understanding of this step in planning a unit.

Facilitator's Notes

The *Framework* identifies seven crosscutting concepts. Their purpose is to help students deepen their understanding of the disciplinary core ideas (NRC 2012, pp. 2 and 8), and develop a coherent and scientifically based view of the world (p. 83). Erickson (2007) would call these crosscutting concepts macroconcepts or integrating concepts because they can transfer across many different disciplines. For example, students should be able to transfer the concept of patterns to patterns in math, patterns in science, and patterns in art. In *How People Learn: Brain, Mind, Experience, and School,* Bransford, Brown, and Cocking (2000) explain that "knowledge is organized around important ideas or concepts that suggest that curricula should also be organized in ways that lead to conceptual understanding" (p. 42). Organizing the *NGSS* around crosscutting concepts will help students conceptually understand science.

In *Essential Questions: Opening Doors to Student Understanding,* McTighe and Wiggins (2013) propose that the role of education is to develop and deepen students' understanding of ideas and processes. Without truly understanding, students will be unable to transfer their science learning to other learning and critical thinking. The authors discuss the idea that essential questions point toward important transferrable ideas that are critical to understand; they can be used to not only stimulate student thinking and inquiry, but to lead to deeper understanding. These essential questions "serve as doorways or lenses through which learners can better see and explore concepts, themes, theories, issues, and problems that reside within the content" (McTighe and Wiggins 2013, pp. 4–5). This activity is designed to demonstrate one process for designing essential questions or "what do we want to learn" using the *NGSS* crosscutting concepts.

Essential questions should align with big ideas. "Essential questions frame ongoing and important inquiries about a big idea" (McTighe and Wiggins 2013, p. 73). The questions should set up genuine and relevant inquiry into big ideas and make meaningful connections with prior learning and personal experiences. Essential questions suggest inquiry and should be concepts in the form of questions. They are used to set the focus for the unit. The crosscutting concepts in *NGSS* are concepts that may be used to make these connections.

Teaching from concepts and not facts is a difficult idea for some teachers to wrap their brain around. It is important that teachers have time to discuss and try multiple examples. They need to grapple with the "so what" of why they teach particular content. What do they really want the students to know? If you see that many are struggling with this concept, we suggest having participants take a copy of one standards page and practice writing essential questions and big ideas from that page in groups. If this concept is going to be new to your participants, plan for this additional opportunity to explore essential questions and big ideas.

Materials

- "What Do We Want to Learn? Essential Questions" handout (p. 143)

- Chart paper and markers (one paper per group of three to four)

- Sticky notes (to provide feedback on the gallery walk)

Procedure

Introduction (10 minutes): Activate prior knowledge about essential questions by asking participants to brainstorm around two different prompts:

1. What is the purpose of an essential question?

2. Identify characteristics of a good essential question.

Provide each table or small group with chart paper and have them identify a participant who will record notes. The recorder should divide the chart paper in half lengthwise. Provide participants with the first prompt (below) and ask them to brainstorm a list of words and short phrases that address the prompt. After a few minutes, provide participants with the second prompt and repeat the brainstorming.

Presentation (15 minutes): This process will work best in groups of three to four. Groups can be mixed or grade-level teams. Use the Facilitator's Notes to provide participants with an overview of the crosscutting concepts and essential questions.

Explore (20 minutes): During this step, participants should be given time to explore the relationships between crosscutting concepts, essential questions, and big ideas. Participants are asked to write essential questions and big ideas for the last two examples in their small groups.

Each group should write their essential questions and big ideas on the chart paper and post their examples for the entire group. Participants should be allowed time to conduct a gallery walk and view the examples. Provide sticky notes for participants to add ideas or suggestions on the examples.

If this is a new activity for your participants, you may want to bring the group back together after page 1 of the handout and discuss participant responses to the questions. This will help everyone have a better idea of the criteria to write an essential question using the crosscutting concepts from *NGSS*. Then, allow groups to try and write their own examples.

Debrief (15 minutes): Ask a few groups to present their examples to the large group. Foster cross talk between groups by asking them to describe the differences in the questions and the big ideas presented by the different groups. Discuss whether some questions are better than others. Also, pose the idea of essential questions that are not driven by the crosscutting concept but by the disciplinary core idea. Do the participants see other essential questions that could be written?

Close the activity by providing participants with a few minutes to reflect (in writing) on what they have learned about essential questions.

Next Steps

If this is your participants' first time using essential questions and big ideas, you may want to allow for additional practice. If your participants are ready to move on, continue to work on the individual pieces for the unit planner as in Activity 20. Once you have introduced the entire key components of the unit planner, your participants will be ready to begin unit planning.

ACTIVITY 18

What Do We Want to Learn?
Essential Questions

Disciplinary core idea	Crosscutting concept	What do we want to learn?	Big idea
3-PS2 Motion and Stability: Forces and Interactions			
The patterns of an object's motion in various situations can be observed and measured; when the past motion exhibits a regular pattern, future motion can be predicted from it.	**Patterns** Patterns of change can be used to make predictions.	Why would we want to predict future patterns of change?	Patterns of change can be used to make predictions of future patterns.
HS-LS4 Biological Evolution: Unity and Diversity			
Genetic information provides evidence of evolution. DNA sequences vary among species, but there are many overlaps; in fact, the ongoing branching that produces multiple lines of descent can be inferred by comparing the DNA sequences of different organisms.	**Patterns** Different patterns may be observed at each of the scales at which a system is studied and can provide evidence for causality in explanations of phenomena.	How might patterns be used to support evolution and common ancestry?	Patterns in genetic information provide evidence of evolution.

1. What is the relationship between the crosscutting concept, the disciplinary core idea, and "What do we want to learn?"

ACTIVITY 18

What Do We Want to Learn? Essential Questions

2. What is the relationship between the crosscutting concept, the disciplinary core idea, and the big idea?

3. What is the relationship between "What do we want to learn?" and the big idea?

4. Are there other "what do we want to learn" questions or different big ideas that could be written for each of these standards?

Now you try it: With a partner complete the curriculum mapping document.

Disciplinary core idea	Crosscutting concept	What do we want to learn?	Big idea
4-ESS3 Earth and Human Activity			
Energy and fuels that humans use are derived from natural sources, and their use affects the environment in multiple ways. Some resources are renewable over time, and others are not.	**Cause and Effect** Cause and effect relationships are routinely identified and used to explain change.		
MS-PS3 Energy			
When two objects interact, each one exerts a force on the other that can cause energy to be transferred to or from the object.	**Systems and System Models** Models can be used to represent systems and their interactions—such as inputs, processes, and outputs—and energy and matter flows within systems.		

ACTIVITY 19

Developing Performance Assessments

Approximate Length

75 minutes

Objectives

During this activity, participants will

- learn what performance assessment is,

- understand the process for writing performance assessment tasks using *NGSS*, and

- develop performance assessment tasks using the performance expectations from *NGSS*.

Vocabulary

- performance expectations

- performance assessment

- performance assessment task

Evidence of Learning

- Concept map of performance assessment

- Graphic organizer "Features of Performance Assessment Tasks"

- Graphic organizer "Turning Performance Expectations into Performance Tasks/ Assessments"

At a Glance

This activity allows participants to explore developing performance assessment tasks using the performance expectations from *NGSS*. Performance assessment tasks demonstrate that students can transfer what they have learned to new situations or challenges. It should be noted however that this activity is just an introduction to performance assessment tasks. The focus of this activity is to help participants understand how to use the performance expectations to begin designing performance assessments. If participants are new to writing performance assessments, additional professional development should be conducted in this area prior to implementation in the classroom.

Facilitator's Notes

Performance expectations in *NGSS* are used to describe what students should be able to do at the end of instruction. Each performance expectation is tied directly to a practice, a disciplinary core idea, and a crosscutting concept from the foundation box. "Performance expectations are the right way to integrate the three dimensions. It provides specificity for educators, but it also sets the tone for how science instruction should look in classrooms" (*NGSS* Executive Summary, pp. 3–4).

Performance assessments are activities that require students to use what they have learned in new ways or situations to demonstrate their learning. "Performance assessments involve students in activities that require them actually to demonstrate performance of certain skills or to create products that demonstrate mastery of certain standards of quality" (Stiggins and Chappuis 2012, p. 138).

It is important to note that the performance expectations help align learning and assessment. When assessments are based on performance expectations, there is an increased chance that they measure what they are supposed to measure. This raises the validity of the assessment.

When looking at the steps to designing a performance assessment task, participants should identify the following pieces:

- A focus for the task: In this case, the performance expectation.

- Context: This is the background knowledge; it gives the role of the student and the overall problem or challenge.

- Audience: The audience for the task.

- What: The requirements of the task and what the final product or performance will be.

- Scoring: The criteria for the task.

As a facilitator you want to make sure that participants arrive at these important steps when they are doing either Group Work Activity 1 or Group Work Activity 2. If groups miss one of the components, you should add it to the discussion during the activity debrief.

Depending on time available and your audience's comfort level with performance assessment, you may want to only facilitate Group Work Activity 1 or Group Work Activity 2. Both activities explore the components of a performance assessment. Each activity uses a different strategy to arrive at the same end result.

Materials

- "Example Performance Assessment Tasks" handout (p. 148–149)

- "Features of Performance Assessment Tasks" handout (p. 150)

- "Turning Performance Expectations Into Performance Tasks/Assessments (Blank)" handout (p. 151)

- "Turning Performance Expectations Into Performance Tasks/Assessments (Scripted)" handout (p. 152)

- Standards for appropriate grade level (For example, we used grade 5 Matter and Its Interactions for our completed template.)

- Chart paper and markers (per group of four to five) with the question "What are features of performance assessment?" written in the middle and circled

Procedure

Set-up: Participants should be organized into small groups prior to starting this activity. When possible, participants should be in grade-level groups. Each group should have markers and a large sheet of chart paper.

Introduction (5 minutes): Provide a brief introduction to the activity. You do not want to explain what performance expectations or performance assessments are at this time. Participants should understand that the next activity is silent. Ask them to number off from 1 to 5 in their group. Tell the groups that you are going to call off a number at the end of the activity, and the person with that number will summarize the group work. This encourages everyone to be engaged.

Group Work Activity 1 (15 minutes): Remind participants that group work will be silent. The facilitator and the groups will not speak until time is called. You should typically allow seven to eight minutes for silent mapping. Assign groups the task of silently making a concept map on the chart paper. They should add keywords or phrases around the context of "What are features of performance assessment?" As the facilitator, feel free to walk around and add connection lines or additional questions that you think might move a map into deeper understanding. At the end of seven to eight minutes, call time. Next, direct participants to discuss aloud for three minutes and determine their group's definition of performance assessment and the key pieces that are needed. Call out a number from 1 to 5 and ask each person with that number to explain his or her group's summary of performance assessment. Record the information presented in the summaries, adding new information from each group. As a whole group, determine your organization's working definition of performance assessment.

Group Work Activity 2 (15 minutes): Pass out the handouts "Characteristics of Performance Assessment Tasks" and "Example Performance Assessment Tasks." Instruct groups to review the performance tasks and look for characteristics and features of each task. Discuss as a group and develop a checklist for common features of a performance task.

Group Work (20–30 minutes): Pass out the completed template and the fifth-grade standard. Allow groups to discuss. If the groups have completed Group Work Activity 1, ask them to look for the features that they identified in the previous activity. Pass out an appropriate grade-level standard to each group and a blank template. Ask groups to generate ideas to complete the template. Each group should complete one template.

Debrief (5–10 minutes): Ask a few of the groups to present their performance assessment to the large group. Foster cross talk between groups by asking each one to describe the process for writing performance assessments using *NGSS*. Have groups explore which pieces are necessary to write a quality performance assessment task.

Next Steps

After this activity, you should assess your participants' understanding of performance assessment. Depending on your participants' level of understanding, you may want to conduct additional professional development on assessment.

ACTIVITY 19

Example Performance Assessment Tasks

1. You have been selected to be the student representative on the school senate. The first item the students have asked you to present is to change the lunch menu in the school cafeteria. Students want the food to be tasty and trendy. You know that unless the new suggested meals meet the U.S. Food and Drug Administration (USDA) food pyramid recommendations the school senate will not vote to change the current offerings. You must design a five-day lunch menu that meets the USDA food pyramid recommendations, yet is tasty enough for students. Then, write a proposal to the school senate explaining why the new menu should be considered.

2. You are an expert in the field of energy conservation. Your consultants have come to you and said that the head of a major electricity provider has claimed there is no need to conserve energy because there is plenty available. Your job is to write a letter to the editor of the local newspaper disputing this claim. You will also design a presentation to present at the next city council's meeting that refutes this electrical company's claim and demonstrates why energy conservation is important.

3. You are planning to go to the park with your friends for the afternoon. Since it is a hot day, you want to serve drinks that will stay cool. When you check out your cupboards at home you discover that you have the choice of three different drink containers: paper cups, Styrofoam cups, and metal cups. You do not know which type of container would keep your drinks the coolest. Design a test to determine which type of cup will keep your drinks the coolest. You have three thermometers, ice, water, and your three types of cups to use in your test. Perform your test. Which type of cups should you take to the park?

4. The local natural history museum has asked your team to build a display for the museum. They are looking for a display to demonstrate the formation of different landforms from your state. Your team is being asked to plan, build, and present a model that can demonstrate a minimum of three different landforms. You will also need to provide background information on the landforms and areas in the state where each may be explored. The museum will feature the best team's display in their Earth science section.

5. You are a world renowned biogeneticist. You have been asked to speak on a panel defending genetically modified organisms. You will need to provide peer-reviewed scientific evidence to support your statements and claims. You will evaluate the validity of the scientific claims made by both proponents and opponents of using genetically modified organisms for food. You will prepare a written report sharing your findings with your panel.

ACTIVITY 19

Features of Performance Assessment Tasks

Directions: Review the sample performance assessment tasks. Individually, look at the features of each task. What features do the tasks have in common? List these in the box below.

> **Common features of vignettes**

Discuss common features of vignettes as a group. Make a group list of common features of performance assessment tasks.

NATIONAL SCIENCE TEACHERS ASSOCIATION

ACTIVITY 19

Turning Performance Expectations Into Performance Tasks/Assessments (Blank)

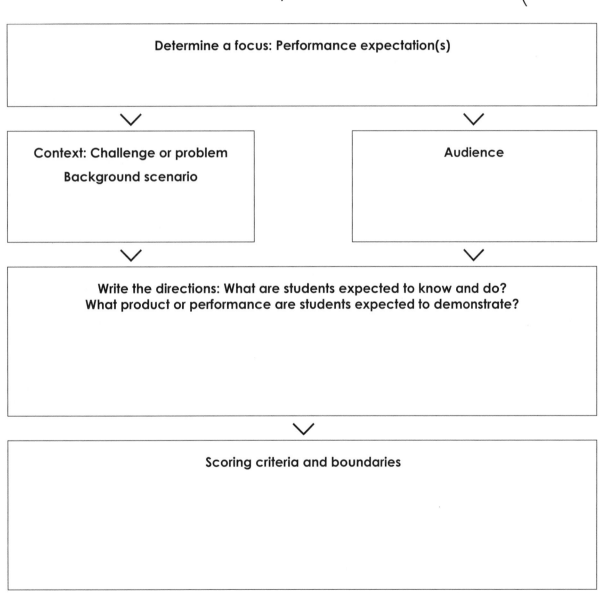

Determine a focus: Performance expectation(s)

Context: Challenge or problem
Background scenario

Audience

Write the directions: What are students expected to know and do?
What product or performance are students expected to demonstrate?

Scoring criteria and boundaries

ACTIVITY 19

Turning Performance Expectations Into Performance Tasks/Assessments (Scripted)

Sample 5th Grade Matter and Its Interactions

Determine a focus: Performance expectation(s)

5-PS1-3. Make observations and measurements to identify materials based on their properties

5-PS1-4. Conduct an investigation to determine whether the mixing of two or more substances results in new substances.

⌄ ⌄

Context: Challenge or problem
Background scenario

You are a hematologist, an expert in the study of blood. An animal's blood has been brought to you to examine and determine what medicine should be given to the animal to save its life.

Audience
The imaginary doctor at the clinic.

⌄ ⌄

Write the directions: What students are expected to know and do? What product or performance are students expected to demonstrate?

The animal has been sick with flu-like symptoms for two days. The doctor tells you that he thinks the animal is suffering from pH-itis. He has been given 3 vials of medicine to treat the animal but does not know which to administer. Administering the wrong medicine could make the animal sicker. The doctor also tells you that normal animal blood for this species should have a pH of 7.0. The doctor gives you a vial of the animal's blood and the 3 vials of medicine. Your job as the blood expert is to design and conduct a test to prove which medicine will cure the animal's case of pH-itis. You will provide a written report with your analysis of the animal's blood and each of the three vials of medicine. You will also make a recommendation on which medicine will cure the animal.

⌄

Scoring criteria and boundaries

- Makes observations and measurements to identify materials based on their properties.

- Conducts an investigation to determine whether the mixing of two or more substances results in new substances. Assessment boundary: Assessment does not include density or distinguishing mass and weight.

NATIONAL SCIENCE TEACHERS ASSOCIATION

ACTIVITY 20

From Standards to Units

Approximate Length

80 minutes

Objectives

During this activity, participants will

- explore the connections between *NGSS* and unit planning,

- design a unit plan based on one *NGSS* standards page, and

- compare and contrast different unit plans for similarities and differences.

Vocabulary

- unit planner

- essential questions

- big ideas

- performance expectations

Evidence of Learning

- Unit planner template

- Feedback form "Feedback on Unit Planning"

At a Glance

This activity builds on the previous curriculum activities. It is an introductory activity that allows participants to explore the connections between

NGSS and unit plans. This activity uses a step-by-step approach to developing unit plans. Participants are walked through writing big ideas, essential questions, what students should be able to know and do, and how they will know what students have learned. This activity will not look at planning learning experiences and instructional strategies to meet the learning needs of all students or many of the other parts of a full unit plan.

Facilitator's Notes

A unit of study can be defined as a series of specific lessons, learning experiences, and related assessments based on designated standards and related supporting standards from a topical, skills-based, or thematic focus that may last anywhere from two to six weeks (Ainsworth 2013, p. 68).

Science units are typically either topical or thematic in nature. Topical units focus on a specific piece of science. For example, topical units in science may be liquids and solids, force and motion, or electricity. Thematic units are broader and may show connections to other disciplines or other topics within the same discipline. Examples of thematic units are ecosystems, life cycles, and water.

It is important to allow participants adequate time to break down the individual parts of the unit plan. Have teachers talk through the process and work collaboratively. Collaboration can help participants clarify questions and issues.

Creating quality science units takes time and hard work. Educators need to understand the individual components and how they fit together. This activity is an introductory activity and does not include all pieces of a completed unit planner. There have been additional items included on this planner to help participants begin to see the integration of *NGSS* components and other disciplines. Most states and districts have additional essential components in their maps. In order to be successful with writing unit plans, educators will require practice with writing units, discussion and collaboration on exemplar units, and meaningful feedback on their written units.

If there is additional time available, it is very helpful to have groups complete a second unit planner on a standard that they will be teaching. Working in collaboration, in a safe environment, helps to reinforce the skills that have been introduced in writing a quality unit plan.

Materials

- Access to one standards page per participant

- "Basic Elements of a Unit Planner" (p. 155), "Unit Planner Template (Blank)" (p. 156), "Annotated Unit Planner Template" (p. 159), and "Completed Unit Planner Template" (p. 162) handouts

Procedure

Introduction (5 minutes): Have participants in grade-level groups of three to four participants. Mixed groups will also work if necessary. Give all participants the same standards page, the "Basic Elements of a Unit Planner" handout, and a blank "Unit Planner Template." Introduce the activity to the whole group. Explain that you are going to allow groups to explore how to write unit plans using the three documents.

Group Work (30 minutes): Give groups 30 minutes to complete a basic unit planner. Groups should have already participated in Activities 19 and 20 and have a grasp of how to begin exploring the unit planner. Encourage groups to use the criteria on the "Basic Elements of a Unit Planner"

handout to focus their work. Remind groups to consider the level of understanding and to review the verbs used on the standards page.

Share (15 minutes): Ask groups to pair up and share their work. Groups should look for similarities and differences in their unit plans. Allow enough time for groups to share their thinking and their decisions on what to put on the unit planner. We have found this to be an important process, so do not rush this step.

Discussion (30 minutes): After the group work, discuss the process together and go through each component as a whole group. You want to use this as a coaching opportunity to have participants think about the components and the quality of each component on their unit planner. Your role as the facilitator is to not only provide information but to ask probing questions that will help participants revise their unit planner. Use the criteria on the "Basic Elements of a Unit Plan" handout to ask questions, and encourage groups to refine their thoughts and ideas.

Debrief (5 minutes): Have participants complete the "Feedback on Unit Planning" form. These should be turned in to the facilitator at the end of the professional development session.

Next Steps

Participants will likely need additional professional development on writing quality unit plans. Districts will also have to decide what they will require in a unit plan and develop the template for educators to use when planning their units. Use the feedback forms provided here to determine participants' needs and future professional development.

ACTIVITY 20
Basic Elements of a Unit Planner

What do we want to learn?

- What provocative question(s) will foster inquiry, understanding, and transfer learning?
- What question can you use to connect this unit to a crosscutting concept?
- Focus on "Why is this important?"
- Written in question form.

Big ideas: I want students to understand ...

- What scientific explanations or models are critical for student understanding of the content?
- So what? Who cares?
- What is most important for students to understand about this topic?
- Will students remember this concept five years from now? ten years from now?

Do: I want students to be able to ...

- What scientific practices will we explicitly focus on in this unit?
- What key skills will students develop as a result of this unit?
- Written as verb phrases.

Know: What are the basics?

- What vocabulary, formulas or other facts do students need to know in order to understand the big ideas?

Technology and the nature of science:

- What connections are there between the unit content and technology or nature of science?

How will I reinforce or build literacy or mathematics skills?

- What *Common Core State Standards* will be taught and assessed in this unit?

How will I know what students have learned?
Performance expectation:

- Does the performance expectation require students to show their understanding in an observable way?
- Does it make students' thinking visible?
- Are there criteria, and are the criteria relevant to the big ideas for the unit?
- Include multiple types of assessment to give a more accurate picture of learning.

What learning experiences will encourage student engagement in the essential questions?

- How shall we teach for understanding?
- Incorporate different learning styles.
- Hands-on and engaging.
- Consider using a learning cycle or BSCS 5E model for planning your instructional sequence.

ACTIVITY 20

Unit Planner Template
(Blank)

Title:	Grade:	Time:
Discipline(s):		

Unit of study:	**Goals to be addressed:** *State content standards, district goals, departmental objectives, student outcomes*
	Common Core State Standards, ELA
	Common Core State Standards, Mathematics

What do we want to learn?

ACTIVITY 20

Unit Planner Template (Blank)

Big ideas: I want students to understand …	Do: I want students to be able to …

Know: What are the basics?	Technology and the nature of science:

How will I reinforce or build literacy or mathematics skills?

How will I know what students have learned?

Performance expectation:

Activity:

Other evidence:

What learning experiences will encourage student engagement in the essential questions (what we want to learn)?

ACTIVITY 20

Annotated Unit Planner Template

Title:	**Grade:**	**Time:**
Discipline(s):		

Unit of study:	**Goals to be addressed:** *State content standards, district goals, departmental objectives, student outcomes*
	Common Core State Standards, ELA
	Common Core State Standards, Mathematics

What do we want to learn?

What provocative question(s) will foster inquiry, understanding, and transfer learning?

What question can you use to connect this unit to a crosscutting concept?

Focus on "Why is this important?"

Written in question form.

Big ideas: I want students to understand ...	**Do: I want students to be able to ...**
What scientific explanations or models are critical for student understanding of the content? So what? Who cares? What is most important for students to understand about this topic?	What scientific practices will we explicitly focus on in this unit? What key knowledge and skills will students develop as a result of this unit? Written as verb phrases.
Know: What are the basics?	**Technology and the Nature of Science:**
What vocabulary, formulas or other facts do students need to know in order to understand the big ideas?	*What connections are there between the unit content and technology or nature of science?*

How will I reinforce or build literacy or mathematics skills?

ACTIVITY 20

Annotated Unit Planner Template

How will I know what students have learned?

Performance expectation:

Does the performance expectation require students to show their understanding in an observable way?

Does it make students' thinking visible?

Are there criteria and are the criteria relevant to the big ideas for the unit?

Activity:

Other evidence:
Include multiple types of learning to give a more accurate picture of learning.

What learning experiences will encourage student engagement in the essential questions?

How shall we teach for understanding?

Incorporate different learning styles.

Hands-on and engaging.

Consider using a learning cycle or BSCS 5E model for planning your instructional sequence.

ACTIVITY 20

Completed Unit Planner Template

Title: Cells to Organisms: Relationships in Structure and Function
Grade: Middle School   Time: 4 weeks
Discipline: Science

Unit of study:	Goals to be addressed: *State content standards, district goals, departmental objectives, student outcomes*
This middle school unit of study will focus on the generalization that all living things are made up of cells. In this study, students will address the idea that cells have structures that are responsible for specialized functions and a group of cells operates a system. Student understandings will also address that the body is a system of multiple interacting subsystems.	LS1.A: Structure and Function • MS-LS1-1 • MS.LS1-2 • MS.LS1-3 *Common Core State Standards, ELA* RI.6.8 Trace and evaluate the argument and specific claims in a text, distinguishing claims that are supported by reasons and evidence from claims that are not. (MS-LS1-3) WHST.6-8.1 Write arguments focused on discipline content. (MS-LS1-3) WHST.6-8.7 Conduct short research projects to answer a question (including a self-generated question), drawing on several sources and generating additional related, focused questions that allow for multiple avenues of exploration. (MS-LS1-1) SL.8.5 Integrate multimedia and visual displays into presentations to clarify information, strengthen claims and evidence, and add interest. (MS-LS1-2) *Common Core State Standards, Mathematics* 6.EE.C.9 Use variables to represent two quantities in a real-world problem that change in relationship to one another; write an equation to express one quantity, thought of as the dependent variable, in terms of the other quantity, thought of as the independent variable. Analyze the relationship between the dependent and independent variables using graphs and tables, and relate these to the equation. (MS-LS1-1), (MS-LS1-2), (MS-LS1-3)

ACTIVITY 20

Completed Unit Planner Template

What do we want to learn? Essential questions

How is cell structure related to cell function?

What is the relationship of cells, multicellular organisms, and other systems in the body?
Is all life made of cells?

Big ideas: I want students to understand ...	Do: I want students to be able to ...
• There is a basic structure and function(s) of the cell. • Cell structure can be visualized, modeled, and used to describe how a cell's function depends on the relationship among its parts. • Multicellular organisms are made up of different types of cells with specialized functions. • The body is a system of multiple interacting subsystems. • All living things are made up of cells, which is the smallest unit that can be said to be alive.	• Observe cells with a microscope • Develop and use a model to describe a cell and the relationship among its parts • Use evidence to support how the body is a system of interacting subsystems

Completed Unit Planner Template

Know: What are the basics?	Technology and the nature of science:
Nucleus, chloroplasts, mitochondria, cell membrane, cell wall Circulatory, excretory, digestive, respiratory, muscular, and nervous system	[From *NGSS* Appendix H, "Understanding the Scientific Enterprise: The Nature of Science"] • Science investigations use a variety of methods and tools to make measurements and observations. • Science knowledge is based upon logical and conceptual connections between evidence and explanations. • Science assumes that objects and events in natural systems occur in consistent patterns that are understandable through measurement and observation.

How will I reinforce or build literacy or mathematics skills?

Students will trace and evaluate the argument and specific claims in a text and use the evidence to support how the body is a system of interacting subsystems.

Students will develop visual models of a cell's structure and function.

Students will read text critically.

Students will use scale to develop a cell model.

How will I know what students have learned?
(We only included one performance expectation in this sample unit; you should include all performance expectations that are going to be assessed in a developed unit.)

Performance expectation:
MS-LS1-2. Develop and use a model to describe the function of a cell as a whole and ways parts of cells contribute to the function.

Activity:
Students will design a cell analogy with each part of the cell structure and each organelle having a corresponding role. Examples might be a town, a factory, a business, a school, or a car. The students will have to explain both the structure and function of each component of the analogy and provide evidence of their reasoning (e.g., lysosome is composting center of a restaurant, breaks down waste, etc.)

Other evidence:
Students will make a diagram/concept map that illustrates the connections between processes that occur in the cell to the same processes that occur in the larger human body (ex. Brain and nucleus are control centers, mitochondria and stomach, etc.)

Quiz:
Class discussion and participation

ACTIVITY 20

Feedback on Unit Planning

Something I learned that I will be applying to my unit planning …	Something specific about unit planning that became clearer to me …
To be successful with unit planning, I will require the following professional development …	**Burning questions I still have …**

9

Connecting
NGSS and the *Common Core State Standards*

Reading furnishes the mind only with materials of knowledge;
it is thinking that makes what we read ours.
—John Locke

The most distinct and beautiful statement of any truth [in science] must
take at last the mathematical form.
—Henry Thoreau

This chapter introduces educators to the connections between the *Next Genera-tion Science Standards* (*NGSS*; NGSS Lead States 2013) and the *Common Core State Standards* (*CCSS*; NGAC and CCSSO 2010) in both mathematics and English language arts (ELA). Research has shown that mathematics and literacy skills are essential to deeper understanding and learning in science (Shanahan, Shanahan, and Misichia 2011). It is important that *CCSS* literacy standards and *CCSS* mathematics standards align with *Next Generation Science Standards* (*NGSS*) because "*NGSS* should always be interpreted and implemented in such a way that they do not outpace or misalign to the grade-by-grade standards in the *CCSS*" (*NGSS*, Appendix M, p. 1). To ensure this, the *NGSS* writing team worked with the *CCSS* teams to identify connections and align the introduction and mastery of content demands (*NGSS* Appendix L and Appendix M).

The activities in this chapter allow participants to begin exploring the connec-tions between *NGSS* and *CCSS*.

ACTIVITY 21

Educators engage in a model activity and use that experience to identify practices, content, and crosscutting concepts from *NGSS* and *CCSS Mathematics* and apply this understanding to possible examples in their own classrooms.

ACTIVITY 22

Educators are introduced to *CCSS ELA* and make connections between *NGSS* and *CCSS ELA* using assessment as the focus.

ACTIVITY 23

Educators examine science articles and provide evidence that the author uses to support the author's position, helping participants to make the connections between *NGSS* and *CCSS* literacy standards.

ACTIVITY 24

Educators explore different styles of text and reading to make connections between the *CCSS* literacy standards and *NGSS*.

ACTIVITY 21

NGSS and the *CCSS Mathematics*

With Amy Parrott

Approximate Length

140 minutes

Objectives

During this activity, participants will

- learn the connections between *NGSS* and *CCSS Mathematics*;

- identify practices, content, and crosscutting concepts from *NGSS* and *CCSS Mathematics*;

- perform a model science activity and identify the science and engineering practices and the *CCSS Mathematics* practices reinforced in the activity; and

- describe a potential activity that integrates standards from *NGSS* and *CCSS Mathematics*.

Vocabulary

- *CCSS*, in mathematics

- *NGSS* connections boxes

- standards of mathematical practice

Evidence of Learning

- Graphic organizer "*NGSS* and *CCSS Mathematics*"

- Graphic organizer "Using the *NGSS* Connections Boxes for *CCSS Mathematics*"

- "Connecting *NGSS* and *CCSS Mathematics* Feedback" form

At a Glance

This activity introduces educators to the connections between *NGSS* and *CCSS Mathematics*. Participants engage in a model activity and use that experience to identify practices, content, and crosscutting concepts from *NGSS* and *CCSS Mathematics*. Participants then apply their understanding of *NGSS*, *CCSS Mathematics*, and the *NGSS* connections boxes to describe a potential activity that integrates standards from both documents.

This activity assumes that participants already have some understanding of how to read the *NGSS* and familiarity with the *NGSS* science and engineering practices and crosscutting concepts.

Facilitator's Notes

The *NGSS* provide connections to the *CCSS Mathematics*. These connections are meant to be starting points for beginning to think about how teachers can integrate science and math while reinforcing concepts in both disciplinary areas. In this activity, participants will be familiarized with the conceptual shifts made by *CCSS Mathematics*, standards for mathematical practice, and the structure of *CCSS Mathematics* (domains, clusters, standards). In addition, participants will identify

how *NGSS* and *CCSS Mathematics* concepts are reinforced in a model activity.

Conceptual Shifts

Shift 1, Focus: *CCSS Mathematics* presents a smaller set of mathematics concepts and allows for a deeper focus on these concepts.

Shift 2, Coherence: The concepts included in *CCSS Mathematics* connect within and across grades so that students have a coherent experience. Early years provide a clear foundation for study in later years.

Shift 3, Fluency: Students are expected to develop fluency—speed and accuracy—with calculations.

Shift 4, Deep Understanding: By focusing on a smaller set of coherent concepts, students develop a deep understanding of math concepts before moving on to more sophisticated concepts.

Shift 5, Application: *CCSS Mathematics* move learning beyond algorithms and simple calculations. Students are expected to be able to apply their math knowledge in authentic situations that may not have a straightforward answer.

Shift 6, Dual Intensity: *CCSS Mathematics* strike a balance between practice and understanding. Students need to have a deep understanding of math concepts and should have sufficient opportunities to practice these concepts.

Standards for Math Practice

The *CCSS Mathematics* include eight standards of mathematical practice.

1. Make sense of problems and persevere in solving them.

2. Reason abstractly and quantitatively.

3. Construct viable arguments and critique the reasoning of others.

4. Model with mathematics.

5. Use appropriate tools strategically.

6. Attend to precision.

7. Look for and make use of structure.

8. Look for and express regularity in repeated reasoning.

The standards for mathematical practice describe expertise and dispositions that should be fostered throughout a students' math education experience. Similar to the integration of content and practices in *NGSS*, an integration of content and practices in math instruction will help students develop mathematical thinking as they approach real-world problems.

Reading *CCSS Mathematics*

The *CCSS Mathematics* are written in grade-specific bands from kindergarten through eighth grade. The standards in each grade level are divided into domains. Related standards are clustered within these domains. In high school, the standards are presented in topics and use the same domain and cluster structure.

For example, there are five domains in grade 5. The text in Figure 9.1 shows one cluster within one domain.

Model Activity Connections

This activity uses a model activity to show how mathematics concepts and practices from the *CCSS Mathematics* can be reinforced during a science activity. A model activity appropriate for elementary teachers and a model activity appropriate for secondary teachers are included in Appendixes 5 and 6. The following discussion describes the *CCSS Mathematics* reinforced in these activities.

Elementary

The content standards met by the model elementary activity are as follows:

FIGURE 9.1

Grade 5: Number and Operations—Fractions (Domain)

Use equivalent fractions as a strategy to add and subtract fractions. (Cluster)

(Standard) *CCSS.Math.Content.5.NF.A.1* Add and subtract fractions with unlike denominators (including mixed numbers) by replacing given fractions with equivalent fractions in such a way as to produce an equivalent sum or difference of fractions with like denominators. For example, $2/3 + 5/4 = 8/12 + 15/12 = 23/12$. (In general, $a/b + c/d = (ad + bc)/bd$.)

(Standard) *CCSS.Math.Content.5.NF.A.2* Solve word problems involving addition and subtraction of fractions referring to the same whole, including cases of unlike denominators, e.g., by using visual fraction models or equations to represent the problem. Use benchmark fractions and number sense of fractions to estimate mentally and assess the reasonableness of answers. For example, recognize an incorrect result $2/5 + 1/2 = 3/7$, by observing that $3/7 < 1/2$.

Standard 2.MD.1 is developed as students measure various distances that a magnet or magnets can push another magnet.

- 2.MD.1: *Measure the length of an object by selecting and using appropriate tools such as rulers, yardsticks, meter sticks, and measuring tapes.*

Students develop standard 2.MD.10 and 3.MD.3 when creating a bar graph of the data (number of magnets pushing versus distance pushed) and answering the related questions.

- 2.MD.10: *Draw a picture graph and a bar graph (with single-unit scale) to represent a data set with up to four categories. Solve simple put together, take-apart, and compare problems using information presented in a bar graph.*

- 3.MD.3: *Draw a scaled picture graph and a scaled bar graph to represent a data set with several categories. Solve one- and two-step "how many more" and "how many less" problems using information presented in scaled bar graphs.*

The formal concept of "average" or "mean" is not defined until middle grades; however, calculating the average of the distance measurements could be classified under standard 3.OA.3.

- 3.OA.3: *Use multiplication and division within 100 to solve word problems in situations involving equal groups, arrays, and measurement quantities.*

Finally, if a teacher chooses to have his or her students use English units instead of or in addition to metric, students can practice standard 3.MD.4.

- 3.MD.4: *Generate measurement data by measuring lengths using rulers marked with halves and fourths of an inch.*

This activity explicitly focuses on at least three standards of mathematical practice. Students are required to reason quantitatively (standard 2) when they make predictions concerning how many paper clips the magnet will hold and how far the magnet(s) will push the other magnet. They are required to use their ruler appropriately (standard 5) and attend to precision (standard 6) while measuring, and they must be precise with the units when reporting their results.

Secondary

The model activity for a secondary-level audience focuses on the following middle school standards from *CCSS Mathematics*:

Students develop standards 5.G.2, 6.NS.8, and 6.SP.4 with the data that they collected.

- 5.G.2: *Represent real world and mathematical problems by graphing points in the first quadrant of the coordinate plane, and interpret coordinate values of points in the context of the situation.*

- 6.NS.8: *Solve real-world and mathematical problems by graphing points in all four quadrants of the coordinate plane.*

- 6.SP.4: *Display numerical data in plots on a number line, including dot plots, histograms, and boxplots.*

When they analyze the graphs and determine which substance conducted heat the best, they are developing the eighth grade standard, 8F.5.

- 8.F.5. *Describe qualitatively the functional relationship between two quantities by analyzing a graph (e.g., where the function is increasing or decreasing, linear or nonlinear). Sketch a graph that exhibits the qualitative features of a function that has been described verbally.*

This activity explicitly focuses on three standards of mathematical practice. While creating the graphs and measuring the temperatures, students must attend to precision (standard 6). When they analyze the data they have collected and determine which substance conducts heat quicker, they are reasoning quantitatively (standard 2) and constructing a viable argument (standard 3).

Materials

- *"NGSS* and *CCSS Mathematics—* Elementary" (p. 175) or *"NGSS* and *CCSS Mathematics—*Middle and High School" (p. 177), "Using the *NGSS* Connections Boxes for *CCSS* Mathematics)" (p. 179), and "Connecting *NGSS* and *CCSS Mathematics* Feedback" (p. 180) handouts

- Access to the *NGSS* and *CCSS Mathematics.* It may be beneficial for participants to have paper copies of the science and engineering practices and the standards for mathematical practice.

- This activity uses one of two model activities. The model activity for use with elementary audiences is "Investigating the Properties of Magnets" from *Activities*

Linking Science with Math, K–4, by John Eichinger, and is located in Appendix 5 (p. 230). The model activity for use with secondary audiences is "Heat Exchange in Air, Water and Soil" from *Activities Linking Science With Math, 5–8*, by John Eichinger, and is located in Appendix 6 (p. 237).

Procedure

This section provides a procedure for elementary audiences (K–5) and a procedure for working with secondary audiences (6–12). If you have a mixed audience, determine which model activity is more engaging for your audience and use the procedure associated with that model activity.

Elementary

Set-up: Before beginning this activity, make sure that you have a working knowledge of *NGSS* Appendix L, "Connections to the *Common Core State Standards* for Mathematics." Prepare materials for the model activity prior to the start of the session. You may also want to separate participants into groups of three to four in advance. The application component of this activity works best if participants are in grade-level teams.

Introduction (15 minutes): Begin by providing participants with an overview of the conceptual shifts made in the *CCSS Mathematics.* These shifts are briefly described in the Facilitator's Notes for this activity. As an alternative, you can show the *Common Core in Mathematics: Overview* video from Engage NY (*www.engageny.org/resource/common-core-in-mathematics-overview*). Discuss with participants how these shifts are similar to those made in the *NGSS* (see Activity 4 [p. 36]). Emphasize that both sets of standards provide a coherent focus on the core ideas that are central to each discipline. In addition, both sets of standards place an emphasis on students being able to use science and mathematics content through the practices.

Model Activity Part I (30 minutes): Engage participants in the first two steps of the elementary model activity. After participants finish these two steps, have them identify the science and engineering practices they engaged in (*NGSS* and *CCSS Mathematics* handouts).

Introducing the Standards for Mathematical Practices (10 minutes): Review the *CCSS Mathematics* standards for mathematical practice. The video, *Importance of Mathematical Practices* from the Hunt Institute (*http://youtu.be/m1rxkW8ucAI*) can be shown to reinforce the importance of focusing on practices. Participants should also review the standards for math practices from the *CCSS* in Mathematics website.

Model Activity Part II (30 minutes): Participants should complete the elementary model activity. After they complete the activity, they should identify the standards for math practices they engaged with (*NGSS* and *CCSS Mathematics* handouts). They should also identify the disciplinary core ideas (*NGSS*), crosscutting concepts (*NGSS*), and content standards (*CCSS Mathematics*) they engaged with in the model activity.

Debrief (15 minutes): Ask participant groups to report out about their discussions in relation to the connections between *NGSS* and *CCSS Mathematics* in the model activity.

Application (30 minutes): Pass out the handout "Using the *NGSS* Connections Boxes for *CCSS Mathematics*." In grade-level groups, participants should identify an *NGSS* standards page to focus on. Briefly draw attention to the connections boxes at the bottom of the standards page and

explain that these are starting points for thinking about how to connect the science and math that students are learning. Implementation of these connections will depend on the sequence of instruction during the school year. Groups should then describe an activity that they could do with their students that integrates both the *NGSS* and *CCSS Mathematics* standards.

Wrap-up (10 minutes): Ask each group to briefly describe the activity and how it reinforces standards from both *NGSS* and *CCSS Mathematics*.

Middle and High School

Set-up: Before beginning this activity, make sure that you have a working knowledge of *NGSS* Appendix L, "Connections to the *Common Core State Standards* for Mathematics." Prepare materials for the model activity prior to the start of the session. You may also want to separate participants into groups of three to four in advance. The application component of this activity works best if participants are in grade-level or content-similar (e.g., biology teachers) teams.

Introduction (25 minutes): Begin by providing participants with an overview of the conceptual shifts made in the *CCSS Mathematics*. These shifts are briefly described in the Facilitator's Notes for this activity. As an alternative, you can show the *Common Core in Mathematics: Overview* video from Engage NY (*www.engageny.org/resource/common-core-in-mathematics-overview*). Discuss with participants how these shifts are similar to those made in the *NGSS* (see Activity 4, p. 36). Emphasize that both sets of standards provide a coherent focus on the core ideas that are central to each discipline. In addition, both sets of standards place an emphasis on students being able to use science and mathematics content through the practices.

Next, introduce participants to the *CCSS Mathematics*. The video *The Importance of Mathematical Practices* from the Hunt Institute (*http://youtu.be/ m1rxkW8ucAI*) can be shown to reinforce the importance of focusing on practices. Participants should also review the standards for math practices from the *CCSS* in mathematics website.

Model Activity Part (45 Minutes): Engage participants in the secondary model activity.

Identification (20 minutes): Distribute the handout "*NGSS* and *CCSS Mathematics*—Middle and High School." Participants should use the *NGSS* and *CCSS Mathematics* to identify the science and engineering practices, standards for mathematics practices, disciplinary core ideas (*NGSS*), content standards (*CCSS Mathematics*), and crosscutting concepts (*NGSS*) used during this activity.

Debrief (15 minutes): Ask participant groups to report their discussions in relation to the connections between *NGSS* and *CCSS Mathematics* in the model activity.

Application (30 minutes): Pass out the handout "Using the *NGSS* Connections Boxes for *CCSS Mathematics*." In grade-level groups, participants should identify an *NGSS* standards page that they can focus on. Briefly draw attention to the connections boxes at the bottom of the standards page and explain that these are starting points for thinking about how to connect the science and math that students are learning. Implementation of these connections will depend on the sequence of instruction during the school year. Groups should then describe an activity that they could do with their students that integrates both the *NGSS* and the *CCSS Mathematics*.

Wrap-up (10 minutes): Ask each group to briefly describe the activity and how it reinforces standards from both *NGSS* and *CCSS Mathematics*. Have all participants complete the "*NGSS* and *CCSS Mathematics* Connections Feedback" form.

Next Steps

This activity begins the process of helping teachers of science understand the connections between *NGSS* and *CCSS Mathematics*. Providing opportunities for same grade-level science and math teachers to use the *NGSS* connections boxes to discuss opportunities for collaboration and curricular areas that can be reinforced in both disciplines can deepen this understanding. Review the feedback forms to look for additional resource and professional development needs of educators.

There are additional resources that can be used to further deepen science teachers' understanding of the *CCSS Mathematics* in Appendix 1 (p. 201).

ACTIVITY 21

NGSS and *CCSS* Mathematics—Elementary

After completing the first part of the activity, create a table to summarize the science and engineering practices you used. Include examples of the "actions" you took during the activity.

Science and engineering practice	How did you engage with this practice?

After completing the second part of the activity, create a table to summarize the standards for mathematical practice that you used. Include examples of the "actions" you took during the activity.

Math practice	How did you engage with this practice?

Use the *NGSS* and the *CCSS* for grade 3 to determine the science and math content that is reinforced during this activity.

Identify the *NGSS* crosscutting concepts related to this activity. How could you make these crosscutting concepts more explicit for your students?

ACTIVITY 21

NGSS and *CCSS Mathematics*— Middle and High School

Create a table to summarize the science and engineering practices that you used. Include examples of the "actions" you took during the activity.

Science and engineering practice	How did you engage with this practice?

Create a table to summarize the standards for mathematical practice that you used. Include examples of the "actions" you took during the activity.

Science and engineering practice	How did you engage with this practice?

Use the *NGSS* and the *CCSS* to determine the science and math content that is reinforced during this activity.

Identify the *NGSS* crosscutting concepts related to this activity. How could you make those concepts more explicit for your students?

ACTIVITY 21

Using the *NGSS* Connections Boxes for *CCSS Mathematics*

Select a standards page from the *NGSS*.

What is the disciplinary core idea?	Which science and engineering practice(s) will you reinforce?

Use the connection box to select a minimum of one content and one practice standard from the *CCSS Mathematics*.

CCSS Mathematics content standard:	*CCSS Mathematics* practice standard:
Describe an activity that you could do with your students that introduces or reinforces the science and math content and practices.	

ACTIVITY 21

Connecting *NGSS* and *CCSS Mathematics* Feedback

Aha—what was something that you learned?	Concerns?

We will need help with ...

Grade level or course_____

NATIONAL SCIENCE TEACHERS ASSOCIATION

ACTIVITY 22

Connecting *NGSS* and *CCSS ELA*

With Mark Bazata

Approximate Length

50 minutes

Objectives

During this activity, participants will

- explore the *CCSS*, English language arts (*CCSS ELA*) and the connections to *NGSS*;

- understand how to use the connections boxes in *NGSS* to identify science and language arts standards connections;

- specify assessments that could be used to assess integrated *NGSS* standards and *CCSS ELA* standards; and

- reflect on how assessment will influence instruction of the standards.

Vocabulary

- *CCSS ELA*
- connections boxes

Evidence of Learning

- Graphic organizer "*NGSS* and *CCSS ELA* Connections"

At a Glance

"Literacy skills are critical to building knowledge in science" (*NGSS*, Appendix M). This activity is designed to help participants see the connections between the *NGSS* and the *CCSS ELA*. Participants are asked to align *NGSS* and *CCSS ELA* standards, then explore possible assessments to demonstrate student learning.

Facilitator's Notes

Reading in science requires students to gather information from a variety of sources, including graphs, charts, and scientific articles that they will only have access to in the science classroom. In addition, writing and speaking in science is different than in other disciplines. The *NGSS* writers were explicit with the connection to the *CCSS ELA* by including them with every disciplinary core idea, science and engineering practice, and grade band. For more information on this connection, please see *NGSS* Appendix M.

This activity is focused on connecting the *CCSS* with the *NGSS* through the use of assessment. Although there is a philosophical connection between both sets of standards, it is only when teachers see how both can be used in the classroom that they will change their pedagogy. Fragmentation in learning is happening too often in education today. This, along with too many standards to address, causes teachers struggle with what is most important and how to teach everything. Connecting standards and disciplines helps teachers see the big picture and increases learning opportunities for students. These connections between *NGSS* and *CCSS* have been made explicit in the standards documents. *NGSS* have listed the *CCSS ELA* standards that connect with the science standard(s) at the bottom of each standards page. *CCSS ELA* grades K–5 in science are integrated in the K–5 *CCSS* reading standards. Grades 6–12 are documented in the section English Language Arts Standards for Science and Technical Subjects in the *CCSS ELA*.

How teachers organize ideas and learning experiences makes a difference in how deeply students understand a concept or skill. Using *NGSS* and *CCSS ELA* allows teachers to scaffold learning. "Understanding requires drawing connections and seeing how new ideas are related to those already learned—how they are alike and different" (Darling-Hammond et al. 2008, p. 198). It is important that teachers not only make the connections in the assessment, but in the instruction that leads up to the assessment.

Materials

- Access to the *NGSS*

- Access to *CCSS ELA* (*www.corestandards.org*)

- "*NGSS* and *CCSS ELA* Connections" handout (p. 184)

Procedure

Set-up (15 minutes): This activity works best with participants in grade-level pairs. Use the Facilitator's Notes and an example of *NGSS* (Middle School PS.1 works well) to provide participants with an overview of connecting the *NGSS* with the *CCSS ELA*. Discuss the alignment between the science standard and the *CCSS ELA* standard. Brainstorm examples of assessments that could be used to demonstrate evidence of learning of these standards. What would the criterion be for the assessment? Discuss how criteria for the assessment can be found directly in the standard. Assessment boundaries from the performance expectations are also helpful in setting criteria. Chart examples on large sheets of paper.

During this brainstorming session, model good questioning skills. As assessment suggestions are offered, focus on key questions such as "Is there a solid connection between the *NGSS* standard and the *CCSS* standard? Is the assessment aligned with both standards? Is the assessment also assessing additional standards that have not been addressed?"

Explore (15 minutes): Have each participant pair up and pick one of the *NGSS* disciplinary core ideas for a grade they currently teach. Each participant should review the *NGSS* as well as the *CCSS ELA* standards identified in the *NGSS* connections box. Each pair should then pick one *NGSS* standard, match that with a *CCSS ELA* standard, and write each standard in the appropriate boxes on the graphic organizer. Finally, brainstorm a list of assessments that could be used to assess both the *NGSS* and the *CCSS* standard(s). Have pairs reflect on how each assessment example will influence instruction of the standards.

Share (15 minutes): For this stage, pairs join up to form foursomes. Each pair takes a turn discussing the listed assessments and then takes a few minutes to add any the other pair can think of.

Debrief (5 minutes): Have each foursome pick one example and share with the large group.

Next Steps

As a follow-up activity, you may want participants to take an existing lesson and align it to a standard from *NGSS* and a standard from *CCSS ELA*. Ask them to reflect on what changes may be necessary in instruction and assessment for the existing lesson to fully meet the performance expectation of the standard. As teachers continue to create new assessments, challenge them to align and score those assessments not only to the *NGSS* but to also consider how the *CCSS* fit in as well.

Instead of having participants align one of their own lessons, you can also introduce them to the Toshiba/NSTA ExploraVision competition. This competition challenges students in grades K–12 to envision how current technology will evolve over the next 20 years and how that technology could be used in the future to solve a problem. The competition requires students to conduct background research, synthesize information from a variety of sources, make predictions about the future, and apply these predictions to solving a societal problem. The task is rich with connections to both *NGSS* and *CCSS ELA*. Participants could review winning entries and the competition rules to develop an understanding of how it aligns to the *CCSS ELA* standards. More information about ExploraVision can be found at *www. exploravision.org*.

ACTIVITY 22

NGSS and *CCSS ELA* Connections

NGSS	CCSS ELA

↘ ↙

Possible connecting assessments

∨

Criteria for assessment

ACTIVITY 23
NGSS and *CCSS ELA*: Connecting Through the Practices

With Mark Bazata

Approximate Length

45–50 minutes

Objectives

During this activity, participants will

- understand what it means to "read like a scientist,"

- understand the relationship between literacy and science,

- identify the argument that an author is making in an article, and

- list evidence that supports an argument.

Vocabulary

- knowledge claims

- argument

- evidence

- explanation

- disciplinary literacy

- "read like a scientist"

Evidence of Learning

- Graphic organizer "Using Evidence to Support an Argument"

At a Glance

This activity asks participants to examine science articles and provide evidence that the author uses to support his or her position. The debriefing part of this exercise demonstrates the relationship between the science and engineering practices and the *CCSS ELA* by having participants read like a scientist and supporting claims with evidence.

Facilitator's Notes

The *NGSS* specifically aligned the standards to the *CCSS ELA* to demonstrate the relationship between science and literacy. This alignment situates literacy as an integral part of content and "literacy within the discipline becomes the goal of disciplinary literacy" (Moje 2008). Fang and Schleppegrell (2008) argue that content area teachers are best suited to teach reading in their respective disciplines because of their knowledge of the content and implicit knowledge of the structure and language of their discipline. Therefore, science teachers are best suited to teach reading of science in science class.

The key to this activity is to see that *CCSS* disciplinary literacy standards are not the only way to get to the science content but do relate well to the science and engineering practices. This activity allows participants to connect the science and engineering practice of engaging in argument from evidence and the *CCSS* disciplinary literacy standards.

When scientists have been questioned about how they read scientific material, the majority have responded that they read while asking questions and thinking about the evidence for what they are reading (Shanahan and Shanahan 2012). Grant and Lapp (2011) would add that asking critical questions is part of being scientifically literate. Critical readers ask questions such as "Why?" and "How?" as they look at scientific texts.

Students should not only be able to ask critical questions but must be able to support the answers to these questions with evidence. The study of science and engineering should produce a sense of the process of argument necessary for advancing and defending a new idea or an explanation of a phenomenon and the norms for conducting such arguments. In that spirit, students should argue for the explanations they construct, defend their interpretations of the associated data, and advocate for the designs they propose (NRC 2012, p. 73). The construction of knowledge claims in science occurs through argumentation in which scientists debate and justify claims using evidence (Driver, Newton, and Osborne 2000). When children engage in such a process and support each other in high quality argument the interaction between the personal and the social dimensions promotes reflexivity, appropriation, and the development of knowledge, beliefs, and values (Vygotsky 1978).

Thinking like a scientist also means reading like a scientist. These skills are important to developing scientific thinking and reasoning strategies that students will be able to use throughout their lives. Teaching students about science requires engaging them in the disciplinary practice of science with a focus on constructing reliable knowledge claims that provide explanatory accounts of nature (Ford and Forman 2006).

So in other words, disciplinary literacy in science is learning science content by focusing on the way reading, writing, speaking and listening, and language are used in the discipline. It is being able to read, write, listen, speak, and think in a way that is meaningful within the context of science. The *CCSS* for Literacy in Science and Technical Subjects and the *NGSS* science and engineering practices connect these areas and provide the "what" that students need to know and be able to do.

Materials

- Two articles taking two different sides of a scientific issue per participant (e.g., whether NASA should be government funded) (We often use articles from *www.procon.org* and *www.beep.ac.uk/content/77.0.html*)

- "Using Evidence to Support an Argument" handout (p. 188)

Procedure

Set-up (15 minutes): Use the Facilitator's Notes to provide participants with an overview of using evidence to support an argument (as in the conclusion of a lab report).

Explore (20 minutes): Give each participant one of the articles on the scientific topic, dividing the two articles evenly between the participants. Ask them to identify the argument that the author is making and list the evidence for his or her argument.

Share (10 minutes): Have each participant pair up with one who read the other article. Go over the evidence from both articles. As a pair, based on the evidence provided, decide which author made the stronger case. Have participants star the evidence boxes that were most important in their decisions.

Debrief (5 minutes): Go through each pair and count up the number of groups that decided on one side or the other for the topic. Have a brief discussion about why it is important to support your arguments in science with evidence.

Next Steps

Have the participants think about a topic that they cover that has a controversial theme. Plan with them how to incorporate an activity that would require students to either find or make an argument and then support it with evidence that they collect.

ACTIVITY 23

Using Evidence to Support an Argument

Name of article:

Argument that the author is making:	
Evidence to support the argument:	**Reasoning or explanation for how it supports the argument:**

ACTIVITY 24

NGSS and *CCSS ELA*: Disciplinary Literacy

With Mark Bazata

Approximate Length

60 minutes

Objectives

During this activity, participants will

- identify the different types of texts that are commonly used with science students, and

- identify strategies to instruct science reading.

Vocabulary

- content area literacy

- disciplinary literacy

- signal words

- connection boxes

Evidence of Learning

- Discussion questions

At a Glance

This activity introduces participants to disciplinary literacy and the connections to *NGSS*. Participants are given the opportunity to explore different styles of text and reading.

Facilitator's Notes

Reading science is a different skill than reading other text. In the past, teachers have been instructed in content reading and the idea of content literacy as "the ability to use reading and writing to learn subject matter in a given discipline" (Vacca and Vacca 2002, p. 15). The strategies were applied across disciplines. For example, paraphrasing is a useful strategy for reading science. However, this strategy would be equally useful in reading other text; it is not specific to science. Shanahan and Shanahan (2008) have proposed disciplinary literacy as the tool that should be used to instruct literacy in different disciplines that is specific to that discipline. "Content area literacy focuses on study skills that can be used to help students learn from subject-matter-specific texts. Disciplinary literacy, in contrast, is an emphasis on the knowledge and abilities possessed by those who create, communicate, and use knowledge within the disciplines" (Shanahan and Shanahan 2012, p. 8).

An example of this would be in the teaching of vocabulary. In content reading, teachers would instruct vocabulary with varied methods. They might use sorting words, puzzles, graphic organizers, a Frayer model, using context, and so on. However, the same strategies would be useful for all disciplines. When you look at science vocabulary, these strategies are not always effective. Many times, students cannot determine the meaning of science vocabulary words by using the context of the reading. If you examine science vocabulary words it becomes evident that many have Greek and Latin roots. For science vocabulary words, it is very helpful for students to know the meaning of these roots. This is an example of a science-specific strategy.

Science text is also organized differently. When reading an excerpt in science writing, the main ideas do not necessarily appear in the expected places. Students have been taught to look for the topic sentence or main idea, which will help them to understand difficult text. Also, many times there are not hierarchy words that explicitly describe relationships. The text is absent of words such as first, another, next, and for example. Readers are taught to look for these signal words. Without signal words, readers must infer which sentences contain new ideas and which are related to prior sentences. For many students, this is a skill that needs to be instructed and practiced.

For students to read and understand informational text, instruction should be specific to the literacy expectations of the discipline. *NGSS* and *CCSS ELA* have made these connections and convey the disciplinary literacy principle that science has a specific approach to literacy knowing and learning.

Materials

- Five large charts, each with one of the following labels: Literature: Stories; Literature: Dramas; Literature: Poetry; Informational Texts: Literary Nonfiction; Informational Texts: Nonfiction

- Sticky notes

- "Disciplinary Literacy and Science" (p. 192), "Cheeks" (p. 193), "The Method of Eratosthenes" (p. 194), "Science Texts" (p. 195), and "Connecting *NGSS* and *CCSS ELA* Feedback" (p. 196) handouts.

Procedure

Introduction (5 minutes): We suggest using grade-level groupings, but mixed groups would also be successful. Briefly discuss the connections between *NGSS* and *CCSS ELA*. If participants

are not aware of where to find standards, show where the *CCSS ELA* are located in the *NGSS* document (in the bottom boxes at the end of each standards page). Also, show the grade 6–12 disciplinary literacy standards in the *CCSS ELA* document (*www.corestandards.org*).

Group Work (15 minutes): Have participants find a partner or join a small group of participants who teach the same grade level. Then, have each group brainstorm a list of texts that are commonly used with students by science educators. On sticky notes, each participant should write down five texts that they typically use—one per note. Next, have participants place their sticky notes in the correct chart (Literature: Stories; Literature: Dramas; Literature: Poetry; Informational Texts: Literary Nonfiction; Informational Texts: Nonfiction).

Discussion and Reflection (10 minutes): Reflect on the placement of the sticky notes on the charts. Where are most of the notes? Pass out the "Disciplinary Literacy and Science" handout. Allow participants a few minutes to read and discuss with their small group. Suggest that groups reflect on the number of minutes that students read weekly in their classroom and the types of text they expose students to weekly.

Group Work (20 minutes): Pass out the "Cheeks" and "The Method of Eratosthenes" handouts. Place participants in groups of three to four participants. Have participants review the two readings. Post discussion questions at the front of the room for each group. What do they notice? What would the teacher's role be if they were using each in a sixth grade classroom? How would vocabulary be taught? Would the reading need to be scaffolded? What instructional strategies would be used with each of the readings?

Debrief (10 minutes): As a group, debrief using the question, "what is disciplinary literacy and how might they connect *NGSS* and disciplinary literacy in their classrooms?" Have participants complete the feedback form. (This feedback form could be used after any of the three *NGSS* and *CCSS ELA* activities.)

Next Steps

Many science educators struggle with the skills to teach disciplinary literacy in their classrooms. Reading like a scientist is different than literary reading. Science teachers, because of their knowledge of the content, are best suited to teach science disciplinary literacy. It is important that educators be provided professional development in strategies to use to instruct disciplinary literacy in their classroom. Review the completed feedback sheets for direction on future professional development. If this is an area that has not been addressed in your school or district, this should be one of your next steps.

ACTIVITY 24

Disciplinary Literacy and Science

Range of text

- We need to start by broadening out our definition of text.
- Texts can be spoken, written, or visual—listened to, read, or viewed.
- A text is any communication—spoken, written, or visual—involving language.
- In an increasingly visual and online world, students need to be able to interpret and create texts that combine words, images, and sound in order to make meaning of texts that no longer read in one clear linear direction.
- Authors select particular text types depending upon their topic, audience, and purpose.
- Texts are published in a variety of digital and print sources, e.g., podcasts, websites, blogs/ microblogs, movies/videos, exhibits, live performances, newspapers, journals, magazines, anthologies, billboards, and fliers.

Source: Adapted from Wisconsin Department of Instruction Form DL-B 2012.

Recommended distribution of the types of reading by grade level

Grade	Literary	Information
4	50%	50%
8	45%	55%
12	30%	70%

Source: National Assessment Governing Board. 2008. Reading Framework for the 2009 National Assessment of Educational Progress. Washington, DC: U.S. Government Printing Office.

Reading in science

- The main idea of a paragraph does not always appear in the first sentence.
- Scientific text does not always contain signal words (first, next, etc). Therefore, students may need to identify which ideas are new and determine the relationships between ideas.
- Generally, in contrast to ordinary language, science is not written in emotional or interpersonal style.
- Usually the text is vocabulary rich, requiring either front loading of text or reading with a dictionary in hand.

ACTIVITY 24

Cheeks

Cheeks looked out from her nest of leaves, high in the oak tree above the Anderson family's backyard. It was early morning and the fog lay like a cotton quilt on the valley. Cheeks stretched her beautiful gray, furry body and looked about the nest. She felt the warm August morning air, fluffed up her big gray bushy tail and shook it. Cheeks was named by the Andersons since she always seemed to have her cheeks full of acorns as she wandered and scurried about the yard.

"I have work to do today!" she thought and imagined the fat acorns to be gathered and stored for the coming of the cold times.

Now the tough part for Cheeks was not gathering the fruits of the oak trees. There were plenty of trees and more than enough acorns for all of the gray squirrels who lived about the yard. No, the problem was finding them later on when the air was cold and the white stuff might be covering the lawn. Cheeks had a very good smeller and could sometimes smell the acorns she had buried earlier. But not always. She needed a way to remember where she had dug the holes and buried the acorns. Cheeks also had a very small memory and the yard was very big. Remembering all of these holes she had dug was too much for her little brain.

The Sun had by now risen in the east and Cheeks scurried down the tree to begin gathering and eating. She also had to make herself fat so that she would be warm and not hungry on long cold days and nights when there might be little to eat.

"What to do ... what to do?" she thought as she wiggled and waved her tail. Then she saw it! A dark patch on the lawn. It was where the Sun did not shine. It had a shape and two ends. One end started where the tree trunk met the ground. The other end was lying on the ground a little ways from the trunk. "I know," she thought. "I'll bury my acorn out here in the yard, at the end of the dark shape and in the cold times, I'll just come back here and dig it up!!! Brilliant Cheeks," she thought to herself and began to gather and dig.

On the next day she tried another dark shape and did the same thing. Then she ran about for weeks and gathered acorns to put in the ground. She was set for the cold times for sure!!

Winter arrived and Cheeks went to dig up her acorns. She looked for the dark end of the shape…

Source: Konicek-Moran, R. 2011. Yet more everyday science mysteries: Stories for inquiry-based science teaching. Arlington, VA: NSTA Press, p. XIII

ACTIVITY 24
The Method of Eratosthenes

The method of Eratosthenes (circa 273–194 BCE) to estimate the Earth's girth consists of measuring the distance separating two distant locations along the north-south line together with their angular separation, determined by measuring the shadows cast by sticks at the two locations (Sarton 1952; Ferguson 1999). Eratosthenes was the director of the Great Library of Alexandria, the greatest library of antiquity. From this enviable position he had access to the most current knowledge in geography, math, and every scientific project of importance. Born in the Greek colony of Cyrene in northern Africa, he received the best education available at the time and became a well-respected scholar in a wide variety of fields, including math and geography. He developed a method to make maps using lines of reference (the precursors of latitude and longitude) and assigning coordinates to geographic places.

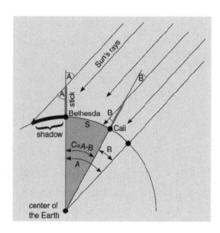

Earth's circumference = S x (360°/C)

To illustrate Eratosthenes' observations, we made a preliminary demonstration in class by shining light on models of a flat and spherical Earth with two sticks of the same size inserted perpendicularly to the surface of each. Whereas the shadows projected on a flat Earth are equal, those on the sphere have different lengths, depending on their location.

Measuring the size of the Earth is a suitable way to begin the development of a precise model of the Earth and the solar system, consistent with the recommendations of the *National Science Education Standards* (Agan and Sneider 2004). The Earth is modeled as a sphere; to connect it with the measurements, we used a circle representing the cross section of the Earth that passes through the two measurement locations. The geometry, as depicted in Figure 1, is defined by this circle and a wedge-shaped section (or slice) as shown. The number of slices in the circle is 360° divided by the angle of the slice (C), and the circumference of the Earth is the distance (S) multiplied by the number of slices (Webb and Bustin 1988).

Source: Torres, S., and J. L. Powers. 2009. Science sampler: Eratosthenes visits middle school: Assessing the ability of students to work with models of the Earth. *Science Scope* 32 (8): 47–48.

ACTIVITY 24

Science Texts

Literature

Novels

Short stories

Poetry

Drama

Movies

Informational: Narrative Nonfiction

Autobiographies

Biographies

Memoirs

Informational: Expository

Newspaper articles

Lectures

Technical journal articles

Textbooks

Magazine articles

Instructional videos

How-to guides

Informational: Persuasive

Editorials

Blog posts

Persuasive speeches

Debates

Informational: Data

Charts

Graphs

Infographics

Data tables

Informational: Visual

Models

Photos

Video footage

Sculpture

Paintings

Illustrations

ACTIVITY 24

Connecting *NGSS* and *CCSS ELA* Feedback

Aha—what was something that you learned?	Concerns?

We will need help with ...

Grade level or course_____

NATIONAL SCIENCE TEACHERS ASSOCIATION

REFERENCES

Ainsworth, L. 2010. *Rigorous curriculum design: How to create curricular units of study that align standards, instruction, and assessment.* Englewood, CO: The Leadership and Learning Center.

Bransford, J. D., A. L. Brown, and R. R. Cocking, eds. 2000. *How people learn: Brain, mind, experience, and school.* Washington, DC: National Academies Press.

CAST, Inc. 2012. *National Center on Universal Design for Learning. www.udlcenter.org.*

Council for Exceptional Children (CEC). 2005. *Universal design for learning: A guide for teachers and education professionals.* Pearson and CEC: Arlington, VA.

Crowther, D. T., N. G. Lederman, and J. S. Lederman. 2009. Understanding the true meaning of the nature of science. *Science and Children* 29 (3): 50–52.

Darling-Hammond, L., B. Barron, P. D. Pearson, A. H. Schoenfeld, E. K. Stages, T. D. Zimmerman, G. Cervetti, and J. Tilson. 2008. *Powerful learning: What we know about teaching for understanding.* San Francisco, CA: Jossey-Bass.

Drago-Severson, E. 2007. Helping teachers learn: Principals as professional development leaders. *Teachers College Record* 109 (1): 70–125.

Driver, R., P. Newton, and J. Osborne. 2000. Establishing the norms of scientific argumentation in classrooms. *Science Education* 84 (3): 287–312.

Duschl, R. A., H. A. Schweingruber, and A. W Shouse. 2007. *Taking science to school: Learning and teaching science in grades K–8.* Washington, DC: National Academies Press.

Erickson, H. L. 2007. *Concept-based curriculum and instruction for the thinking classroom.* New York: Corwin Press.

Fabing, H., and R. Marr, eds. 1944. *Fischerisms: Being a sheaf of sundry and divers utterances culled from the lectures of Martin H. Fischer, professor of physiology in the University of Cincinnati.* Springfield, IL: C. Thomas.

Fang, Z., and M. Schleppegrell. 2008. *Reading in secondary content areas: A language-based pedagogy.* Ann Arbor, MI: University of Michigan Press.

Fogarty, R., and B. Pete. 2010. Professional learning 101: A syllabus of seven protocols. *Phi Delta Kappan* 91 (4): 32–34.

Grant, M., and D. Lapp. 2011. Teaching science literacy. *Educational Leadership* 68 (6). *www.ascd.org/ publications/educational-leadership/mar11/vol68/ num06/Teaching-Science-Literacy.aspx.*

Guskey, T. R. 2000. *Evaluating professional development.* Thousand Oaks, CA: Corwin Press.

Hunzicker, J. 2011. Effective professional development for teachers: A checklist. *Professional Development in Education* 37 (2): 177–179.

References

Jorgenson, O. 2000. The need for more ethnic teachers: Addressing the critical shortage in American public schools. *Teachers College Record. www.tcrecord.org* (ID No. 10551).

Konicek-Moran, R. 2001. *Yet more everyday science mysteries: Stories for inquiry-based science teaching.* Arlington, VA: NSTA Press.

Kucer, S. B., C. Silva, and E. L. Delgado-Larocco. 1995. *Curricular conversations: Themes in multilingual and monolingual classrooms.* York, ME: Stenhouse Publishers.

Leithwood, K., K. S. Louis, S. Anderson, and K. Wahlstrom. 2004. *How leadership influences student learning.* New York: The Wallace Foundation.

Little, J. W. 1993. Teachers' professional development in a climate of education reform. *Education Evaluation and Policy Analysis* 15 (2): 129–151.

Locke, J. *On the conduct of understanding* (written 1697, published posthumously 1706), collected in Works. 5th ed. 1751. Vol. 3, 387.

Martin-Kniep, G. 1999. *Capturing the wisdom of practice: Professional portfolios for educators.* Alexandria, VA: ASCD.

McLaughlin, M. W., and D. D. Marsh. 1978. Staff development and school change. *Teachers College Record* 80 (1): 70–94.

McTighe, J., and G. Wiggins 2005. *Understanding by design.* 2nd ed. Alexandria, VA: ASCD.

McTighe, J., and G. Wiggin. 2013. *Essential questions: Opening doors to student learning.* Alexandria, VA: ASCD.

McTighe, J., and G. Wiggins, G. 2005. *Understanding by design.* 2nd ed. Alexandria, VA: ASCD.

Meridith, D. 1979. Carl Sagan's cosmic connection and extraterrestrial life-wish. *Science Digest* 85: 37

Moje, E. B. 2008. Foregrounding the disciplines in secondary literacy teaching and learning: A call for change. *Journal of Adolescent and Adult Literacy* 52 (2): 97–106.

National Assessment Governing Board. 2008. *Reading framework for the 2009 National Assessment of Educational Progress.* Washington, DC: U.S. Government Printing Office.

National Commission on Mathematics and Science Teaching (NCMST). 2000. *Before it's too late: A report to the nation from the National Commission on Mathematics and Science Teaching for the 21st Century.* Washington, DC: NCMST

National Governors Association Center for Best Practices and Council of Chief State School Officers (NGAC and CCSSO). 2010. *Common core state standards.* Washington, DC: NGAC and CCSSO.

National Research Council (NRC). 2000. *Inquiry and the National Science Education Standards: A guide for teaching and learning.* Washington, DC: National Academies Press.

National Research Council (NRC). 2006. *America's lab report: Investigations in high school science*, ed. S. R. Singer, M. L. Hilton, and H. A. Schweingruber. Washington, DC: National Academies Press

National Research Council (NRC). 2012. *A framework for K–12 science education: Practices, crosscutting concepts, and core ideas.* Washington, DC: National Academies Press.

NGSS Lead States. 2013. *Next Generation Science Standards: For states, by states.* Washington, DC: National Academies Press. *www.nextgenscience.org/ next-generation-science-standards.*

Peterson, J. M., and M. M. Hittie. 2010. *Inclusive teaching: The journey toward effective schools for all learners.* 2nd ed. Boston: Pearson.

Senge, P. M., N. H. Cambron-McCabe, T. Lucas, A. Kleiner, J. Dutton, and B. Smith. 2000. *Schools that learn: A fifth discipline fieldbook for educators, parents, and everyone who cares about education.* New York: Doubleday.

Shanahan, T., and C. Shanahan. 2008. Teaching disciplinary literacy to adolescents: Rethinking content area literacy. *Harvard Educational Review* 78 (1): 40–59.

Shanahan, T., and C. Shanahan. 2012. What is disciplinary literacy and why does it matter? *Topics in Language Disorders* 32 (1): 7–18.

Shanahan, C., T. Shanahan, and C. Misichia. 2011. Analysis of expert readers in three disciplines: History, mathematics, and chemistry. *Journal of Literacy Research* 43 (4): 393–429

Short, D., and J. Echevarria. 1999. The Sheltered Instruction Observation Protocol: A tool for teacher-research collaboration and professional development. Education Practice Report, Santa Cruz, CA and Washington DC: Center for Research on Education, Diversity and Excellence. *www.cal.org/resources/digest/sheltered.html.*

Stiggins, R., and J. Chappuis. 2012. *An introduction to student-involved assessment for learning.* 6th ed. Boston: Pearson.

Tai, R. H., P. M. Sadler, and J. J. Mintzes. 2006. Research and teaching: Factors influencing college science success. *Journal of College Science Teaching* 42 (9): 987–1012.

Thoreau, H. D. 1873. *A week on the Concord and Merrimack rivers.* Boston, MA: J. R. Osgood.

Torres, S., and J. L. Powers. 2009. Science sampler: Eratosthenes visits middle school: Assessing the ability of students to work with models of the Earth. *Science Scope* 32 (8): 47–48.

Tyson, N. G. 1994. *Universe down to Earth.* New York: Columbia University Press.

Udvari-Solner, A., R. A. Villa, and J. S. Thousand. 2005. Access to the general education curriculum for all: The universal design process. In *Creating an inclusive school*, ed. R. A. Villa and J. S. Thousand, 134–155. 2nd ed. Alexandria, VA: ASCD.

Udvari-Solner, A. 1995. A process for adapting inclusive classrooms. In *Creating an inclusive school*, ed. R. A. Villa and J. S. Thousand, 110–124. Alexandria, VA: ASCD.

Vacca, R., and J. Vacca. 2002. *Content area reading: Literacy and learning across the curriculum.* 7th ed. Boston: Allyn and Bacon.

Valle, J. W., and D. J. Conner. 2010. *Rethinking disability: A disability studies approach to inclusive practices.* New York: McGraw-Hill.

Vygotsky, L. 1978. *Mind and society.* Cambridge, MA: Harvard University Press.

Wiggins, G., and J. McTighe. 2005. *Understanding by design.* 2nd ed. Alexandria, VA: ASCD.

Wiggins, G., and J. McTighe. 2011. *Understanding by design guide to creating high-quality units.* Alexandria, VA: ASCD.

Willis, J. A. 2006. Preserve the child in every learner. *Kappa Delta Pi Record* 44 (1): 33–37.

Whitmore, K., and C. Crowell. 1994. *Inventing a classroom.* York, ME: Stenhouse.

APPENDIXES

APPENDIX 1

Resources

These resources provide additional depth to the activities in this book.

Chapter 1

- Pratt, H. 2013. *The NSTA reader's guide to the* Next Generation Science Standards. Arlington, VA: NSTA Press.

- Pratt, H. 2013. *The NSTA reader's guide to A framework for K–12 science education: Practices, crosscutting concepts, and core ideas.* 2nd ed. Arlington, VA: NSTA Press.

- Banko, W., M. L. Grant, M. E. Jabot, A. J. McCormak, and T. O'Brien, eds. 2013. *Science for the next generation: Preparing for the new standards.* Arlington, VA: NSTA Press.

Chapter 2

- The 5 Features of Science Inquiry, Edutopia Blog Series (Brunsell):

  - What questions do you have? *www. edutopia.org/blog/five-features-science-inquiry*

  - How do you know? *www.edutopia.org/ blog/teaching-science-inquiry-based*

  - Designing science inquiry, *www. edutopia.org/blog/science-inquiry-claim-evidence-reasoning-eric-brunsell*

- Bozeman Science—*Next Generation Science Standards* videos by Paul Andersen: *www. bozemanscience.com/next-generation-science-standards/*

- Brunsell, E. 2012. *Integrating engineering and science in your classroom.* Arlington, VA: NSTA Press.

- Moyer, R., and S. Everett. 2012. *Everyday engineering: Putting the E in STEM teaching and learning.* Arlington, VA: NSTA Press.

- University of California Museum of Paleontology's Understanding Science: *http://undsci.berkeley.edu/index.php*

- Rothstein, D., and L. Santana. 2011. *Make just one change: Teach students to ask their own questions.* Cambridge, MA: Harvard Education Press.

Chapter 3

- Fathman, A. K., and D. T. Crowther, eds. 2006. *Science for English language learners: K–12 classroom strategies.* Arlington, VA: NSTA Press.

- Finson, K. D., C. K. Ormsbee, and M. M. Jensen. 2011. *Differentiating science instruction and assessment for learners with special needs, K–8.* Thousand Oaks, CA: Corwin Press.

- Howard, L. A., and E. A. Potts. 2013. *Including students with disabilities in advanced science classes.* Arlington, VA: NSTA Press.

- Linz, E., M. J. Heater, and L. A. Howard. 2011. *Team teaching science: Success for all learners.* Arlington, VA: NSTA Press.

- National Center on Universal Design for Learning: *www.udlcenter.org*

- Special issue on diversity and equity in science education. *Theory Into Practice* 2013 52 (1).

Chapter 4

- Erickson, H. L. 2007. *Concept-based curriculum and instruction for the thinking classroom.* Thousand Oaks, CA: Corwin Press.

- Jacobs, H. H. 2010. *Curriculum 21: Essential education for a changing world.* Alexandria, VA: ASCD.

- Literacy Design Collaborative: *www.literacydesigncollaborative.org/tasks/*

- McTighe, J., and G. Wiggins. 2013. *Essential questions: Opening doors to student learning.* Alexandria, VA: ASCD.

- McTighe, J., and G. Wiggins. 2011. *The understanding by design guide to creating high-quality units.* Alexandria, VA: ASCD.

Chapter 5

Mathematics

- *NGSS* Appendix L: Connections to the *Common Core State Standards* for Mathematics

- A Closer Look at the *Common Core State Standards* for Mathematics—workshop material: *http://educationnorthwest.org/resource/1800*

- Illustrative Mathematics: *www.illustrativemathematics.org*

- Inside Mathematics: *www.insidemathematics.org*

- Common Core Tools: *commoncoretools.me*

- Achieve the Core: *www.achievethecore.org*

English Language Arts

- Achieve the Core: *www.achievethecore.org*

- Jetton, T. L., and C. Shanahan. 2012. *Adolescent literacy in the academic disciplines: General principles and practical strategies.* New York: Guilford Press.

- Literacy Design Collaborative: *www.literacydesigncollaborative.org*

APPENDIX 2

Model Activity (K–5)

Scientific Background Information

A force is generally defined as a push or pull. Forces are required to set an object in motion, changeits direction, or stop its motion. Newton's first law of motion tells us that an object at rest stays at rest and an object in motion stays in motion in a straight line at a constant speed unless acted upon by an external force. The only way to move something that is still is to apply a force, to push or pull it. Likewise, the only way to stop an object that is in motion is to apply a force, to push or pullin the opposite direction of the motion. At this grade level it is not necessary for students to state Newton's first law of motion. What you want students to understand is that forces are required to start, stop, and change the direction of an object's motion.

Two forces that are always present on Earth are gravity and friction. Gravity pulls objects toward Earth. When an object is tossed into the air, gravity's pull slows the object even as it is rising in the air. At this point, gravity is pulling in the opposite direction of the object's motion. When the object begins to fall back to Earth, the force of gravity is acting in the same direction of the motion.

This results in an increased velocity in a downward direction. We see the force of gravity in action when an object slides down a ramp. Gravity pulls on the object as it travels down the ramp. At this grade level, what you want your students to understand is that gravity is a force that pulls objects toward Earth.

Friction is a force that acts in the opposite direction to an object's motion. Friction slows objects, but it also works to hold objects in place. To set an object in motion, the force has to be great enough to overcome the weight of the object and the force of friction that holds the object in place. Frictional forces act in the opposite direction of an object's motion, slowing it down. Friction exists everywhere. It occurs when two objects are in contact with each other. The smoother the object, the lower the frictional force. The big idea here for students is that friction is present everywhere, holds objects in place, and slows the motion of objects.

When a force sets a stationary object in motion, the object accelerates in the direction of the force. The greater the force, the greater the acceleration. This is one part of Newton's second law of motion. The mass of an object in relation to force is the other part. Newton's second law tells us that acceleration is directly proportional to force and inversely proportional to mass. This means that:

- If the same force is applied to two objects with different masses, the objects will accelerate at different rates. The object with the larger mass will have a lower rate of acceleration. The one with the smaller mass will have a greater rate of acceleration.

- If different forces are applied to two objects with the same mass, the objects will have different rates of acceleration. A larger force will result in greater acceleration than a smaller force.

It is not necessary for students to state Newton's second law. What you want them to come away with is an understanding that larger forces cause objects to move faster than do smaller forces and that heavy objects require more force to move than lighter objects.

Materials

- Masking tape

- Curling sliders, four per team (You can use furniture sliders, which are available at home improvement stores and discount department stores.)

- Bumpers (You need objects that students can place in the classroom curling playing area to redirect the sliders; board erasers work well.)

- "Changing Force I" data sheet (p. 206)

Explore

In this phase students will work in small groups to investigate how forces speed up, slow down, and change the direction of objects in motion.

Changing Direction

In this part of the explore phase, students demonstrate that a slider travels in a straight path unless it collides with a bumper.

1. Students will first confirm that the slider travels in a straight path if no external force acts on it. Put the slider on a smooth surface (e.g., the floor or a tabletop). Give it a push with a moderate amount of force. The slider should travel in a fairly straight path.

2. Model for the class how to diagram the path the slider follows by drawing it on the board. Students can replicate the diagram

in their science notebooks. Students will create additional diagrams later in the unit.

3. Discuss with the students the types of information the diagram illustrates.

4. Use these guiding questions to summarize the activity and transition to the next steps:

   o What do you notice about the path of the slider? (Is it straight, curved, zigzagged?)

   o How might you change the direction of the slider?

   o What would happen if you pushed the slider harder? More gently?

5. Demonstrate that when a slider collides with a bumper, it changes direction. Students will have observed this when they played classroom curling. Demonstrating it here provides an opportunity to model how to diagram the slider's path. To demonstrate:

   a. Set a bumper at an angle in the path of the slider as shown in Figure 9.2.

   b. Diagram the setup on the board.

   c. Ask students to predict what will happen when the slider collides with the bumper.

   d. Launch the slider so that it reflects off the bumper when they collide.

   e. Diagram the path of the slider.

FIGURE 9.2.

Path of the slider

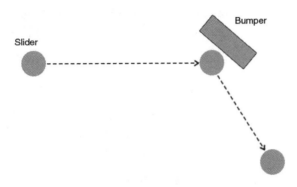

6. Teams will now test the effect of placing bumpers in a slider's path. Each team should test three different bumper positions one at a time, as follows:

   a. Select and diagram the bumper position as demonstrated.

   b. Predict what will happen when the slider collides with the bumper.

   c. Launch the slider so that it reflects off the bumper when they collide.

   d. Diagram the path of the slider.

   e. Conduct two more tests.

As the students explore, ask guiding questions like:

   o What happens to the direction the slider is moving when a bumper is placed in its path?

   o Which bumper position caused the greatest change in the slider's direction?

7. When all teams have finished testing, bring the class back together as a whole group. Ask teams to share and compare their findings.

8. As a class, make a general statement about what happens when an object moving in a straight line experiences an outside force. Write the statement on the board. Students will record the statement in their science notebooks. These questions will help guide students' thinking:

   o How would you describe the path of the slider when there are no bumpers in the way?

   o What happens when the slider runs into the bumper? Does this always happen?

   o Why does the direction of the slider change when it runs into a bumper? (If students are having difficulty drawing the conclusion that the bumper applies a force to the slider, reread page 6 of Exploring Forces and Movement.)

Changing Force I

In this part of the explore phase students will investigate what happens when an increasing force is applied to the slider.

Advance preparation: For each team, cut a piece of string that is approximately 15 cm longer than the width of the desk. Securely tape a snack-size plastic bag to one end of the string, and tape the free end to the top of a slider (see Figure 9.3, p. 206).

1. Draw the "Changing Force I" data table on the board.

FIGURE 9.3

Changing Force I Experimental setup

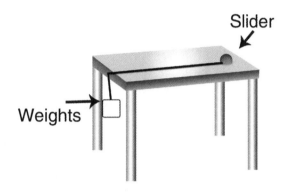

Slider

Weights

2. Demonstrate the process students will use to test the effect that changing the force in the direction of the motion has on the speed of the slider. Put the slider near one edge of a desk and stretch the string across the desk so that the plastic bag hangs over the opposite side as illustrated in Figure 9.3. Put enough weight in the bag so that when it is released it slowly pulls the slider across the desk.

3. Repeat the demonstration using a stopwatch to time how long it takes the slider to reach the other side of the desk. Record the time in the data table that is on the board.

4. Give each group one slider with the string and plastic bag attached, a stopwatch, weights, and the "Changing Force I" data sheet.

5. Tell students to follow the procedure and the demonstration to complete their investigation.

As the students explore, ask guiding questions like:

- How does increasing the weight impact the time it takes for the slider to travel across the table?

- What direction is the weight moving?

- What direction is the slider moving?

- What do you think would happen if the slider were heavier? (If desired, place a weight on top of the slider to test this.)

6. Students now find the mean (average) of the three trials for each test. Caution students to be sure to use the observed rather than the predicted times.

7. Ask students to create a bar graph with time on the vertical axis and number of weights on the horizontal axis. If they are not mathematically ready to calculate the mean, tell them to use the median (middle) time for their graph.

8. After creating the graph, students should share and compare with the whole class.

9. As a class, write a general statement about how increasing the force changes the motion of the slider. Post this statement on chart paper in a prominent place for the remainder of the unit.

This excerpt and the worksheet are reprinted with permission from Fries-Gaither, J., and T. Shiverdecker. 2013. Classroom curling: Exploring forces and motion. In Inquiring scientists, inquiring readers: Using nonfiction to promote science literacy, grades 3–5. pp. 111–131. Arlington, VA: NSTA Press. Figure numbers are from the original book.

APPENDIX 2

Changing Force I

Materials

- Weights
- Slider with weight bag attached
- Stopwatch or watch with second hand

Slider
Weights

Procedure

1. Add enough weight to the bag so that it slowly pullsthe slider across the desk.

2. Record the number of weights in the correct columnon the data table.

3. Conduct three trials and record the time it takes for the slider to be pulled across the desk for each trial.

4. Add more weight to the bag and repeat the process.

Test	Number of weights	Time		
		Trial 1	Trial 2	Trial 3
1				
2				

What do you notice about the time it takes for the slider to be pulled across the desk when the weight is increased?

Based on what you noticed, predict how long it will take for the slider to be pulled across the desk when more weight is added.

Test 3 _____ number of weights _____ predicted time

Test 4 _____ number of weights _____ predicted time

Test	Number of weights	Time		
		Trial 1	Trial 2	Trial 3
1				
2				

The weights are a force pulling on the slider. How does increasing the weight (force) change the motion of the slider?

APPENDIX 3

Model Activity (6–12)

Chapter 5: Now You "Sea" Ice, Now You Don't

Penguin communities shift on the Antarctic Peninsula

Teacher Pages

At a Glance

Increasing air temperatures in the last 50 years have dramatically altered the Antarctic Peninsula ecosystem. In this interdisciplinary inquiry, learners use a cooperative approach to investigate changes in the living and nonliving resources of the Peninsula. The activity stresses the importance of evidence in the formulation of scientific explanations. (Class time: 1–3.5 hours)

"For many, [the Antarctic Peninsula] is the most beautiful part of the Antarctic, unlocked each year by the retreating ice. … It is on this rocky backbone stretching north that most of the continent's wildlife survives. … Almost every patch of accessible bare rock is covered in a penguin colony. Even tiny crags that pierce the mountainsides are used by nesting birds."

—Alastair Fothergill, A Natural History of the Antarctic: Life in the Freezer, 1995

Introduction

At the global level, strong evidence suggests that observed changes in Earth's climate are largely due to human activities (IPCC 2007). At the regional level, the evidence for human-dominated change is sometimes less clear. Scientists have a particularly difficult time explaining warming trends in Antarctica—a region with a relatively short history of scientific observation and a highly variable climate (Clarke et al. 2007). Regardless of the mechanism of warming, however, climate change is having a dramatic impact on Antarctic ecosystems.

By the end of this lesson, students should be able to do the following:

- Graphically represent data.

- Use multiple lines of evidence to generate scientific explanations of ecosystem-level changes on the Antarctic Peninsula.

- Describe ways in which climate change on the Antarctic Peninsula has led to interconnected, ecosystem-level effects.

- Participate in an interdisciplinary scientific investigation, demonstrating the collaborative nature of science.

Warming Climate, Waning Sea Ice

Air temperature data indicate that the western Antarctic Peninsula (Figure 5.4, [p. 212]) has warmed by about 3°C in the last century (Clarke et al. 2007). Although this relatively short-term record is only from a few research stations, other indirect lines of evidence confirm the trend. The most striking of these proxies is a shift in penguin communities. Adélie penguins, which are dependent on sea ice for their survival, are rapidly declining on the Antarctic Peninsula despite a 600-year colonization history. In contrast, chinstrap penguins, which prefer open water, are dramatically increasing (Figure 5.1).

These shifts in penguin populations appear to be the result of a decrease in the amount, timing, and duration of sea ice (Figure 5.2).

Why is sea ice so important to Adélie penguins? First, sea ice is a feeding platform for Adélies. Krill, the primary prey of Adélies on the Peninsula, feed on microorganisms growing on the underside of the

ice (Atkinson et al. 2004). For Adélie penguins, which are relatively slow swimmers, it is easier to find food under the ice than in large stretches of open water (Ainley 2002). Second, sea ice helps control the local climate. Ice keeps the Peninsula cool by reflecting solar radiation back to space. As air temperatures increase and sea ice melts, open water releases heat and amplifies the upward trend in local air temperature (Figure 5.3) (Wadhams 2000). Finally, ice acts as a giant cap on the ocean, limiting evaporation. As sea ice declines, **condensation nuclei** (aerosols that form the core of cloud droplets) and moisture are released into the atmosphere, leading to more snow. This extra snow often does not melt until Adélies have already started nesting; the resulting melt water can kill their eggs (Fraser and Patterson 1997).

Activity Overview

This directed inquiry uses the jigsaw technique, which requires every student within a group to be an active and equal participant for the rest of the group to succeed (Colburn 2003). To begin, students are organized into "Home Groups"

FIGURE 5.1

Adélie and chinstrap penguins.

Adélie penguins (*Pygoscelis adeliae*) breed on the coast of Antarctica and surrounding islands. They are named after the wife of French explorer Jules Sébastien Dumont d'Urville. Adult Adélies stand 70–75 cm tall and weigh up to 5 kg.

Chinstrap penguins (*Pygoscelis antarctica*) are primarily found on the Antarctic Peninsula and in the Scotia Arc, a chain of islands between the tip of South America and the Peninsula. Their name comes from the black band running across their chins. Adult chinstraps stand 71–76 cm tall and weigh up to 5 kg.

Source: Photographs courtesy of Michael Elnitsky.

FIGURE 5.2

Effects of climate change on sea ice, krill, and penguin communities of the Antarctic Peninsula.

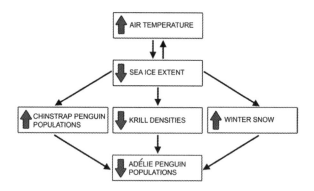

FIGURE 5.3

Melting sea ice amplifies the effects of climate change.

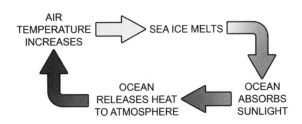

composed of five different specialists. Specialists from each Home Group then reorganize into "Specialist Groups" that contain only one type of scientist (e.g., Group 1 could include all of the Ornithologists and Group 2 all of the Oceanographers). Each Specialist Group receives a piece of the flowchart in Figure 5.2, in the form of a data table. With only a few facts to guide them, the Specialist Groups create graphs from the data tables, brainstorm explanations for patterns in their data, and report results back to their Home Groups. Finally, Home Groups use the expertise of each specialist to reconstruct the entire flowchart (Figure 5.2).

Teaching Notes

Prior Knowledge

Before starting this activity, students should have at least a rudimentary knowledge of Antarctica. You can find a collection of links to our favorite Antarctic websites at *www.units.muohio.edu/cryolab/education/AntarcticLinks.htm*. You can also engage student interest in this inquiry by showing video clips of penguins, which are naturally appealing to students of all ages. We have short

movies of Adélies feeding their young and battling predators on our website at *www.units.muohio.edu/cryolab/education/antarcticbestiary.htm*, and National Geographic has a video called "Rocky Parenting" at *http://news.nationalgeographic.com/news/2006/11/061117-adelie-video.html*.

Materials

- Specialist Fact Sheet (Student Page 5.1 [p. 217]; one for each student, or one overhead for the entire class)

- Temperature data (Figure 5.4 [p. 212]; one overhead for the entire class)

- Data sets for each Specialist Group (Student Pages 5.2–5.6 [pp. 218–222]: Adélie Penguins, Sea Ice, Winter Snow, Chinstrap Penguins, and Krill)

- Specialist Group Report Sheets (Student Page 5.7 [p. 223]; one for each student)

- Sheets of graph paper (one for each student) or computers connected to a printer (one for each Specialist Group)

- Sets of six flowchart cards (one complete set for each Home Group; before the inquiry, you can make flowchart cards by photocopying Figure 5.2 and cutting out each box [i.e., "Air Temperature," "Sea Ice Extent," etc.])

- Paper, markers, and tape for constructing flow charts

Procedure: Graphing and Interpretation

1. Split the class into Home Groups of at least five students each. (Optional: Assign the name of a different real-life research agency to each group. See *www.units.muohio.edu/*

cryolab/education/AntarcticLinks.htm#Nt-nlProg for examples.)

2. Instruct students to read the Specialist Fact Sheets (Student Page 5.1 [p. 217]). Within a Home Group, each student should assume the identity of a different scientist from the list.

3. Introduce yourself: "Welcome! I'm a climatologist with the Palmer Station, Antarctica Long-Term Ecological Research project. In other words, I study long-term patterns in climate. My colleagues and I have tracked changes in air temperatures on the peninsula since 1947. We have observed that although temperature cycles up and down, it has increased overall (show Figure 5.4). We think this is occurring because of an increase of greenhouse gases, but we are unsure of the impacts on the Antarctic ecosystem. Your team's job is to describe the interconnected effects of warming on Antarctica's living and nonliving systems."

4. Direct the specialists to meet with their respective Specialist Groups. Specialist Groups should not interact with one another.

5. Distribute the data sets and Specialist Group Report Sheets (Student Pages 5.2–5.7 [pp. 218–223]) to each Specialist Group. The specialists should graph their data set and interpret the graph.

Procedure: Flowchart and Class Discussion

1. Reconvene the Home Groups.

2. Hand out a complete set of flowchart cards to each group. Each specialist should

FIGURE 5.4

Climatologists: Air temperature data set.

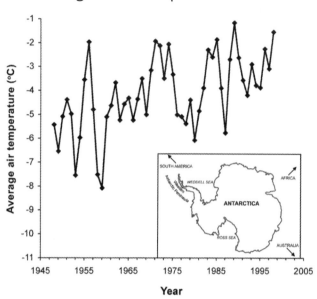

Source: Data compiled from the Palmer Station, Antarctica Long-Term Ecological Research (LTER) data archive. Data from the Palmer LTER archive were supported by the Office of Polar Programs, NSF Grants OPP-9011927, OPP-9632763, and OPP-021782.

make a brief presentation to his or her Home Group approximating the format on the Specialist Group Report Sheet (Student Page 5.7 [p. 223]). Home Groups should then construct their own flowcharts using all of the flowchart cards. Remind the students throughout this process that they should use the weight of evidence to construct the flowcharts. In other words, each idea should be accepted or rejected based on the amount of support it has.

3. Consider these discussion questions during the flowchart process (do this as a class, by Home Group, or as homework for each student):

FIGURE 5.5

Performance rubric.

Student Name:

Criteria	Points	Performance rubric		
		Self	Teacher	Comments
Active participation in the group process..	5			
Appropriate graph is used to display data. All required elements (labels, titles, etc.) are present. Data are graphed accurately.	5			
Data and interpretations from Specialist Groups are clearly communicated to Hope Groups by individual specialists.	10			
Alternative explanations are weighed based on available evidence and prior scientific knowledge.	10			
Conclusions are clearly and logically communicated.	10			
Report Sheet is complete.	5			
TOTAL	**45**			

- How has the ecosystem of the Antarctic Peninsula changed in the last 50 years? What are the most likely explanations for these changes?

- Is there sufficient evidence to support these explanations? Why or why not? What further questions are left unanswered?

- Did your Specialist Group come up with any explanations that you think are not very likely (or not even possible!), based on the complete story presented by your Home Group?

Assessment

To assess student learning, you can use a simple performance rubric that focuses on group work and the nature of science (Figure 5.5). Depending on the unit of study in which this inquiry is used, a variety of specific content standards also may be assessed. In an ecology unit, for example, you could determine student knowledge of interactions between populations and their environments; in an Earth science unit, you could check student understandings about weather and climate.

Modifications

Some students have initial difficulties with the construction and interpretation of flowcharts. Once students have connected their flowchart cards with arrows, it may be useful to have them label each arrow with a verb. For instance:

↓ Sea Ice Extent →[causes]→ ↑Winter Snow Fall

For lower-level students, you can construct a worksheet with a "skeleton" of the worksheet (e.g., the general shape of the flowchart and some of the text within the boxes).

You can shorten this lesson by starting immediately with Specialist Groups, rather than with Home Groups. Another option is to provide pre-made graphs of the data rather than having Specialist Groups create their own.

To make this lesson more open-ended, students may do additional research on the connections between sea ice, krill, and penguins. Note, however, that the majority of resources on this topic are research articles in scientific journals. If you have access to a university library, you might wish to make a classroom file of related journal articles. A more engaging extension would be for students to generate ideas for new research studies that would address questions left unanswered by the current inquiry.

This type of activity could range from asking students to formulate new hypotheses to asking students to write short proposals that include specific research questions and plans to answer those hypotheses.

Conclusion

Many students have trouble comprehending how just a few degrees of atmospheric warming (in this case, 3°C) could make a difference in their lives. The decline of a charismatic species such as the Adélie penguin is an example of how a seemingly minor change in climate can pose a major threat to plants and animals. Beyond the effects of climate change, however, the activity illustrates the multidisciplinary, international, and, above all, cooperative nature of science. We want social teenagers to realize that they do not have to sit alone in a lab to do science.

References

Ainley, D. G. 2002. *The Adélie penguin: Bellwether of climate change.* New York: Columbia University Press.

Atkinson, A., V. Siegel, E. Pakhomov, and P. Rothery. 2004. Longterm decline in krill stock and increase in salps within the Southern Ocean. *Nature* 432(7013): 100–103.

Clarke, A., E. J. Murphy, M. P. Meredith, J. C. King, L. S. Peck, D. K. A. Barnes, and R. C. Smith. 2007. Climate change and the marine ecosystem of the western Antarctic Peninsula. *Philosophical Transactions of the Royal Society B* 362(1477): 149–166.

Colburn, A. 2003. *The lingo of learning: 88 education terms every science teacher should know.* Arlington, VA: NSTA Press.

Fraser, W. R., and D. L. Patterson. 1997. Human disturbance and long-term changes in Adélie penguin populations: A natural experiment at Palmer Station, Antarctic Peninsula. In *Antarctic communities: Species, structure and survival,* eds. B. Battaglia, J. Valencia, and D. W. H. Walton, 445–452. Cambridge: Cambridge University Press.

Intergovernmental Panel on Climate Change (IPCC). 2007. Summary for policymakers. In *Climate change 2007: The physical science basis. Contribution of Working Group I to the Fourth Assessment*

Report of the Intergovernmental Panel on Climate Change, eds. S. Solomon, D. Qin, M. Manning, Z. Chen, M. Marquis, K. B. Averyt, M. Tignor, and H.

L. Miller, 1–18. Cambridge: Cambridge University Press.

Smith, R. C., W. R. Fraser, and S. E. Stammerjohn. 2003. Climate variability and ecological response of the marine ecosystem in the Western Antarctic Peninsula (WAP) region. In *Climate variability and ecosystem response at Long-Term Ecological Research sites*, eds. D. Greenland, D. G. Goodin, and R. C. Smith, 158–173. New York: Oxford University Press.

Wadhams, P. 2000. *Ice in the ocean.* The Netherlands: Gordon and Breach Science Publishers.

Other Recommended Resources

These additional resources were used to create the Student Pages, but are not cited in the text:

Carlini, A. R., N. R. Coria, M. M. Santos, and S. M. Buján. 2005. The effect of chinstrap penguins on the breeding performance of Adélie penguins. *Folia Zoologica* 54(1–2): 147–158.

Forcada, J., P. N. Trathan, K. Reid, E. J. Murphy, and J. P. Croxall. 2006. Contrasting population changes in sympatric penguin species in association with climate warming. *Global Change Biology* 12(3): 411–423.

Fraser, W. R., W. Z. Trivelpiece, D. G. Ainley, and S. G. Trivelpiece. 1992. Increases in Antarctic penguin populations: Reduced competition with whales or a loss of sea ice due to environmental warming? *Polar Biology* 11(8): 525–531.

Lynnes, A. S., K. Reid, and J. P. Croxall. 2004. Diet and reproductive success of Adélie and chinstrap penguins: Linking response of predators to prey population dynamics. *Polar Biology* 27(9): 544–554.

Moline M. A., H. Claustre, T. K. Frazer, O. Schofield, and M. Vernet. 2004. Alteration of the food web along the Antarctic Peninsula in response to a regional warming trend. *Global Change Biology* 10(12): 1973–1980.

Newman, S. J., S. Nicol, D. Ritz, and H. Marchant. 1999. Susceptibility of Antarctic krill (*Euphausia superba* Dana) to ultraviolet radiation. *Polar Biology* 22(1): 50–55.

Nicol, S. 2006. Krill, currents, and sea ice: *Euphausia superba* and its changing environment. *BioScience* 56(2): 111–120.

Parkinson, C. L. 2004. Southern Ocean sea ice and its wider linkages: Insights revealed from models and observations. *Antarctic Science* 16(4): 387–400.

Turner, J., S. R. Colwell, and S. Harangozo. 1997. Variability of precipitation over the coastal western Antarctic Peninsula from synoptic observations. *Journal of Geophysical Research* 102(D12): 13999–14007.

This chapter and the following student pages are reprinted with permission from Constible, J., L. Sandro, and R. E. Lee. 2008. Now you "sea" ice, now you don't: Penguin communities shift on the Antarctic peninsula. In Climate change from pole to pole: Biological investigations, pp. 122–127. Arlington, VA: NSTA Press. Figure numbers are from the original book.

Chapter 5

Now You "Sea" Ice, Now You Don't

Penguin communities shift on the Antarctic Peninsula

Student Pages

Note: Reference List for Students

For more information on references cited in the Chapter 5 Student Pages,
go to teacher references on page 85.

Student Page 5.1

Specialist Fact Sheet

Each Home Group contains five different specialists:

1. *Ornithologist:* A scientist who studies birds. Uses visual surveys (from ship or on land), diet analysis, bird banding, and satellite tracking to collect data on penguins.

2. *Oceanographer:* A scientist who studies the ocean. Uses satellite imagery, underwater sensors, and manual measurements of sea-ice thickness to collect data on sea-ice conditions and ocean temperature.

3. *Meteorologist:* A scientist who studies the weather. Uses automatic weather stations and visual observations of the skies to collect data on precipitation, temperature, and cloud cover.

4. *Marine Ecologist:* A scientist who studies relationships between organisms and their ocean environment. Uses visual surveys, diet analysis, and satellite

   tracking to collect data on a variety of organisms, including penguins.

5. *Fisheries Biologist:* A scientist who studies fish and their prey. Collects data on krill during research vessel cruises.

Student Page 5.2

Ornithologists (Adélie Penguin Data Set)

- Adélie penguins spend their summers on land, where they breed. They spend winters on the outer extent of the sea ice surrounding Antarctica, where they molt their feathers and fatten up.

- Adélies are visual predators, meaning they need enough light to see their prey. Near the outer part of the pack ice, there are only a few hours of daylight in the middle of the winter. There is less sunlight as you go farther south (closer to land).

- On the western Antarctic Peninsula, Adélie penguins mostly eat krill, a shrimplike crustacean.

- Several countries have been heavily harvesting krill since the mid-1960s.

- Adélie penguins need dry, snow-free places to lay their eggs. They use the same nest sites each year and at about the same time every year. Heavy snowfalls during the nesting season can bury adult Adélies and kill their eggs.

- Female Adélies lay two eggs, but usually only one of those eggs results in a fledged chick (fledged chicks have a good chance of maturing into adults). The two most common causes of death of eggs and chicks are abandonment by the parents (if they cannot find enough food) and predation by skuas (hawklike birds).

- In the water, Adélies are eaten mostly by leopard seals and killer whales.

- Adélies can look for food under sea ice because they can hold their breath for a long time. They are not as good at foraging in the open ocean, because they cannot swim very fast.

- Adélie penguins have lived in the western Antarctic Peninsula for at least 644 years.

YEAR	# BREEDING PAIRS OF ADÉLIE PENGUINS
1975	15,202
1979	13,788
1983	13,515
1986	13,180
1987	10,150
1989	12,983
1990	11,554
1991	12,359
1992	12,055
1993	11,964
1994	11,052
1995	11,052
1996	9,228
1997	8,817
1998	8,315
1999	7,707
2000	7,160
2001	6,887
2002	4,059

Source: Data compiled from Smith, Fraser, and Stammerjohn. 2003. Photograph courtesy of Richard E. Lee, Jr.

Student Page 5.3

Oceanographers (Sea Ice Data Set)

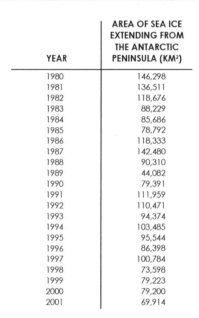

- In August or September (the middle of winter), sea ice covers over 19×106 km^2 of the Southern Ocean (an area larger than Europe). In February (the middle of summer), only 3×106 km^2 of the ocean is covered by sea ice.

- Sea ice keeps the air of the Antarctic region cool by reflecting most of the solar radiation back into space.

- Open water absorbs solar radiation instead of reflecting it and converts it to heat. This heat warms up the atmosphere.

- Sea ice reduces evaporation of the ocean, thus reducing the amount of moisture that is released to the atmosphere.

- As sea ice melts, bacteria and other particles are released into the atmosphere. These particles can form condensation nuclei, which grow into rain or snow.

- Rain helps to stabilize the sea ice by freezing on the surface.

- Sea ice can be broken up by strong winds that last a week or more.

- An icebreaker is a ship with a reinforced bow to break up ice and keep channels open for navigation. Icebreakers were first used in the Antarctic in 1947 and have been commonly used to support scientific research for the last 25 years.

YEAR	AREA OF SEA ICE EXTENDING FROM THE ANTARCTIC PENINSULA (KM²)
1980	146,298
1981	136,511
1982	118,676
1983	88,229
1984	85,686
1985	78,792
1986	118,333
1987	142,480
1988	90,310
1989	44,082
1990	79,391
1991	111,959
1992	110,471
1993	94,374
1994	103,485
1995	95,544
1996	86,398
1997	100,784
1998	73,598
1999	79,223
2000	79,200
2001	69,914

Source: Data compiled from the Palmer Station, Antarctica Long-Term Ecological Research (LTER) data archive. Data from the Palmer LTER archive were supported by the Office of Polar Programs, NSF Grants OPP-9011927, OPP-9632763, and OPP-021782. Photograph courtesy of Marianne Kaput.

Student Page 5.4

Meteorologists (Winter Snow Data Set)

- In the winter, most of the precipitation in the western Antarctic Peninsula occurs as snow. There is an even mix of snow and rain the rest of the year.

- It is difficult to accurately measure the amount of snowfall in the Antarctic because strong winds blow the snow around.

- The Antarctic Peninsula has a relatively warm maritime climate so it gets more rain and snow than the rest of the Antarctic continent.

- Most of the rain and snow in the western Antarctic Peninsula is generated by cyclones from outside the Southern Ocean. Cyclones are areas of low atmospheric pressure and rotating winds.

- When there is less sea ice covering the ocean, there is more evaporation of the ocean and, therefore, more moisture in the atmosphere.

- As sea ice melts, bacteria and other particles are released into the atmosphere. These particles can form condensation nuclei, which grow into rain or snow.

YEAR	% OF PRECIPITATION EVENTS THAT ARE SNOW
1982	49
1983	67
1984	72
1985	67
1986	81
1987	80
1988	69
1989	69
1990	68
1991	72
1992	70
1993	70
1994	83
1995	77
1996	74
1997	81
1998	81
1999	83
2000	77
2001	90
2002	82
2003	76

Source: Data compiled from Antarctic Meteorology Online, British Antarctic Survey (*www.antarctica.ac.uk/met/metlog*). Photograph courtesy of Luke Sandro.

Student Page 5.5

Marine Ecologists (Chinstrap Penguin Data Set)

- Chinstrap penguins breed on land in the spring and summer and spend the rest of the year in open water north of the sea ice. The number of chinstraps that successfully breed is much lower in years when the sea ice does not melt until late in the spring.

- Chinstraps mostly eat krill, a shrimplike crustacean.

- Whalers and sealers overhunted seals and whales, which also eat krill, until the late 1960s.

- Chinstraps hunt primarily in open water because they cannot hold their breath for very long.

- The main predators of chinstraps are skuas (hawklike birds), leopard seals, and killer whales.

- Chinstraps will aggressively displace Adélie penguins from nest sites in order to start their own nests and may compete with Adélies for feeding areas.

- Although chinstrap penguins have occupied the western Antarctic Peninsula for over 600 years, they have become numerous near Palmer Station (one of the three U.S. research stations in Antarctica) only in the last 35 years.

YEAR	# BREEDING PAIRS OF CHINSTRAP PENGUINS
1976	10
1977	42
1983	100
1984	109
1985	150
1989	205
1990	223
1991	164
1992	180
1993	216
1994	205
1995	255
1996	234
1997	250
1998	186
1999	220
2000	325
2001	325
2002	250

Source: Data compiled from Smith, Fraser, and Stammerjohn. 2003. Photograph courtesy of Michael Elnitsky.

Student Page 5.6

Fisheries Biologists (Krill Data Set)

- Krill, a shrimplike crustacean, is a keystone species, meaning it is one of the most important links in the Antarctic food web. All the vertebrate animals in the Antarctic either eat krill or another animal that eats krill.

- Krill eat mostly algae. In the winter, the only place algae can grow is on the underside of sea ice.

- Several countries have been harvesting krill since the mid-1960s.

- Ultraviolet radiation is harmful to krill and can even kill them. Worldwide, ozone depletion is highest over Antarctica.

- Salps, which are small, marine animals that look like blobs of jelly, compete with krill for food resources. As the salt content of the ocean decreases, salps increase and the favorite food species of krill decrease.

YEAR	DENSITY OF KRILL IN THE SOUTHERN OCEAN (# KRILL/M^2)
1982	91
1984	50
1985	41
1987	36
1988	57
1989	15
1990	8
1992	7
1993	22
1994	6
1995	9
1996	31
1997	53
1998	46
1999	4
2000	8
2001	31
2002	8
2003	3

Source: Data compiled from Atkinson et al. 2004.
Photograph courtesy of Richard E. Lee, Jr.

Student Page 5.7

Specialist Group Report Sheet

Name:_____

Specialist Group:_____

In your own words, summarize the general trends or patterns of your data. Attach a graph of your data to the back of this sheet.

List possible explanations for the patterns you are seeing.

With the help of the facts on each data sheet, choose the explanation that you think is most likely. Why do you think that explanation is most likely?

APPENDIX 4

Model Activity for the Nature of Science

If I have seen farther it is because I stand on the shoulders of giants.
—Isaac Newton

Even the greatest of scientists owe their accomplishments to those who came before them. They built on the theories of their predecessors and laid the foundations of future science. This is a concept that I try to communicate to my students. Scientists do not work in isolation, nor do they simply come up from scratch with brilliant ideas fully formed in all details. Scientists learn what has been previously discovered and theorized and then question and test those ideas with new technology and new ways of thinking. Today's scientific theories are the result of a long collaborative process, sometimes over centuries, among many different scientists from various parts of the world.

When I introduce the theory of plate tectonics, I try to show how, even though it is credited to Alfred Wegener (1880–1930), it was in reality a long collaborative process taking place over many decades. My strategy is to place students in the role of geologists attending a major conference: The 23rd Annual Consortium of Geologists. The purpose of this conference is to examine some peculiar new discoveries (see Figures 1–6) and to see if this information can lead to new ideas about the Earth. This lesson is an introduction to the concept of plate tectonics and continental drift. Students have had no prior experience with these subjects. My goal is to have students examine a new series of evidence and try to develop their own hypotheses based on this evidence. If students can look at the same evidence as the early proponents of continental drift and come up with similar conclusions, then they are more likely to retain the concept. I present a series of six pieces of evidence (Figures 1–6), from the shape of the continents to earthquakes and magnetic resonance, in approximately the same order in which these discoveries actually happened. This evidence can be found in a variety of textbooks, as well as resources such as The Story of Science series by Joy Hakim (2004; 2005; 2007) and A Short History of Nearly Everything by Bill Bryson (2004). As students examine the evidence, they are asked questions to try to find connections and to develop ideas on what the evidence seems to indicate, as well as what questions still need answering. The goal is to have students develop the theory of plate tectonics from the same evidence as the original scientists and, because the follow-up discussion and explanation I give to students tells when these pieces of evidence were first discovered, to have students see that this was a very long process. The following is the process, strategy, and materials I use to present this lesson. A general timeline of events is provided as a guide for teachers to see the actual time frame and order of events in the development of plate tectonics.

This entire lesson takes one 60-minute class period. This lesson is used as an introduction to plate tectonics. It follows my unit on evolution and classification, so students are familiar with fossils and change over time.

Timeline of Events

(This timeline is initially for teacher background, but is provided to students as part of a follow-up explanation of the real events in history.)

1906: Alfred Wegener visits Greenland as part of an expedition studying polar circulation. He returns again in 1912.

1911: Wegener finds scientific papers in the University of Marburg library that tell of identical fossils of plants and animals found on opposite sides of the Atlantic Ocean. Current theories speculate the existence of land bridges between the continents, but Wegener notices the peculiar shape of the continents and how they seem to fit together. He begins to doubt conventional wisdom and searches for more evidence.

1911–1915: Through research Wegener discovers that there are many large-scale features such as the Scottish Highlands and Appalachian Mountains, and unique rock strata such as the Santa Catarina system of Brazil and the Karroo system of South Africa, that show amazing similarities across the ocean. He also notes that the found fossils were from climates far different from present-day climates on those continents.

1915: Wegener publishes The Origin of Continents and Oceans. This book outlines his idea of continental drift. It is received with outright hostility from other scientists. The criticism is mainly due to his lack of a mechanismthat could describe how the continents moved.

FIGURE 1

Concept for consideration: Do the shapes of the continental edges indicate any historical significance?

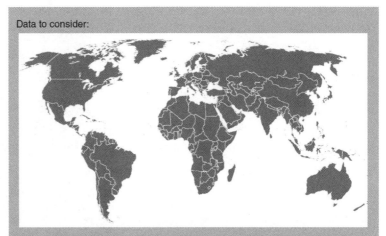

Data to consider:

Examine the outline of the various continents on the map. Are there any patterns, particularly surrounding the Atlantic Ocean? If so, what significance can you make of those patterns?

1930: Alfred Wegener dies during his final expedition to Greenland. He gets lost and freezes to death on the ice.

1940s: World War II brings about an increased interest in studying the oceans now that submarine warfare is common and detailed maps of the ocean floor are needed.

1950s: Studies of the ocean floor and earthquakes lead to detailed maps showing the increasing age of rocks as you travel outward from the mid-Atlantic ridge. Scientists also plot the location of earthquake epicenters around the world and see a distinctive pattern rather than a random distribution of epicenters.

1960s: Increases in technology and further study of seismic waves lead to more detailed information about the Earth's interior. Convection currents in the mantle are discovered and now used as an explanation of the forces involved in movement of the continents. The theory of plate tectonics is now generally accepted.

Part 1—The Setup

As the class begins, I welcome everyone to the 23rd Annual Consor tium of Geologists. I explain that they (students) are all invited guests to this conference and, as leading experts in geology, they are to work in small groups to examine some startling new evidence that has come to light. Our task is to develop some sort of theory to explain these data.

The class is divided into groups of three or four and each group is provided with the first three papers (Figures 1–3) showing the shape of the continents, fossil evidence, and glacier deposits. They are asked to look over these papers and come up with an explanation as to what might have caused all of this to happen. Students can write down their thoughts and modifications to their ideas as we proceed so they have something to refer to during the discussion times that follow. There will also be a final summar y report (see Part 4) to write, and students' notes will help with the writing. After 5–10 minutes, I get everyone's attention and ask the following questions:

1. What is the evidence being presented?

2. What explanations could you come up with to explain this evidence?

FIGURE 2

Concept for consideration: Do the fossils found in different continents indicate any historical significance?

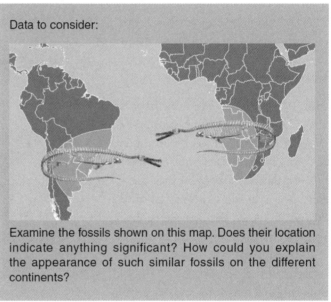

Data to consider:

Examine the fossils shown on this map. Does their location indicate anything significant? How could you explain the appearance of such similar fossils on the different continents?

3. Is this enough evidence to develop a theory?

4. What questions do you still have?

5. What evidence is still needed to develop a solid theory?

I find that most small groups are indeed able to develop the idea that the land masses must have moved from some other location to account for this evidence. Some groups need more assistance and guiding questions than others, but this can be determined as I listen to each group's discussion. The questions help to solidify and clarify the concept.

Part 2

To continue the scenario of the geologists' consortium, I tell students that we are now several weeks into our conference and further new evidence has just been brought to light. I give each group the next two papers (Figures 4 and 5); one is on the age of rocks on either side of the mid-Atlantic ridge and the other is the epicenter location map. Students are asked to examine this new evidence and connect it to the evidence they had previously. What new connections can they find? After 5–10 more minutes of discussing how the new evidence fits in with the original three papers and adding this to their notes and modifying their hypothesis as needed, I ask the next round of questions:

1. What is the new evidence that was presented?

2. Were you able to see any connections between this evidence and what you were given earlier? If so, what did you find?

3. How did this evidence affect your previous ideas? Did it require an entirely new theory or just an adjustment to your old one?

4. What questions still need to be asked?

5. What evidence is still missing?

Students should continue to take notes on the evidence and the process so that they will have something to refer to when they write their final summary (see Part 4).

FIGURE 3

Concept for consideration: Do glacier deposits found in different continents indicate any historical significance?

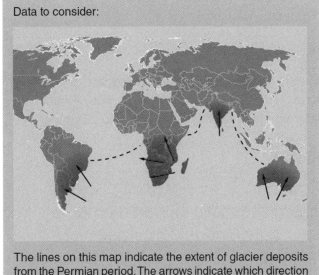

Data to consider:

The lines on this map indicate the extent of glacier deposits from the Permian period. The arrows indicate which direction the glaciers were moving. What does this evidence suggest to you? Remember, glaciers can only form over land, never over water. Also remember that glaciers of this size must be from extremely cold climates.

Part 3

To complete the imaginary scene, I tell students that we are in the final week of the consortium and more evidence has come to our attention. I ask students to examine one last paper (Figure 6) to see if it can finally answer our questions about the Earth. I give each group the final paper showing the convection currents in the Earth's mantle and ask them to add this to the others (Figures 1–5) and examine them as a whole to see what they can determine. After 5–10 minutes, I ask the last round of questions:

1. What is the new evidence?

2. How does this new evidence affect your previous ideas?

3. What theory, or theories, have you been able to develop based on all of the evidence presented?

Part 4

In this final part of the lesson, possibly as a homework assignment, students individually write a brief summary, approximately one page in length,

FIGURE 4

Concept for consideration: Do the age of the rocks found on either side of the mid-Atlantic ridge indicate any historical significance?

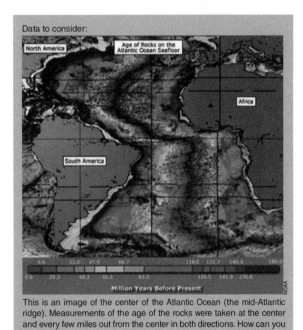

Data to consider:

This is an image of the center of the Atlantic Ocean (the mid-Atlantic ridge). Measurements of the age of the rocks were taken at the center and every few miles out from the center in both directions. How can you explain the age of the rocks seen on either side of the center? Remember, the mid-Atlantic ridge is very active with volcanoes and lava eruptions.

of the evidence that was presented and the hypothesis that their group developed as a result of examining this evidence. This written work should clearly explain how the evidence combines to support their theory. After they turn in their summaries, and as a follow-up the next day, I tell students the story of how the actual theory was developed over many decades (see timeline of events) and how this lesson was developed to simulate that process. I usually do this by reciting the timeline of events as if it were a story, reminding students of the pieces of evidence they examined themselves during the lesson. Students are generally amazed by their ability to look at the same evidence used by the scientists and develop the same ideas. This is a very successful lesson and students find it engaging and thought provoking. They enjoy trying to piece together the puzzle and are very proud of the fact that they too can look at evidence and develop the same ideas as the scientists.

Alternative Approach (Jigsaw)

While using the same scenario of a geologists' convention, I sometimes alter the groups based on dynamics and to vary things a bit for my own sanity. I divide the class into groups of six and then hand each group member a different evidence paper. I then ask all of the group members with the same papers (all fossil papers, or all convection current papers) to sit together and talk in their group about the evidence and see if they can explain what their evidence is showing. Students should again take notes on the discussions so that they can relay the information clearly to their original groups. I then ask the original groups to get back together and share what they have learned. Each group is asked to piece together all of the evidence and develop a theory as to what all of this evidence tells us.

FIGURE 5

Concept for consideration: Do the locations of earthquake epicenters indicate any historical significance?

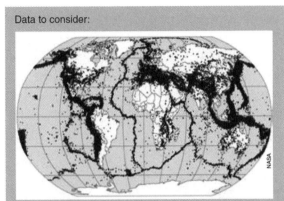

Data to consider:

This map indicates the location of earthquakes over a 30-year time span. Do you see any pattern in the location of the earthquakes? How do you explain the pattern? Why might earthquakes happen in these locations?

I guide the groups as needed, using many of the same questions as shown above. The same written summary is also required at the end.

This introductory lesson leads us to our unit on Earth changes with lessons on seafloor spreading, earthquakes, volcanoes, weathering, erosion, and ocean floor topography. We also tie in lessons on how plate tectonics is connected to other concepts such as historical climate change and biological evolution.

Jim Cronin (jcronin@mail.ccsd.k12.co.us) is a science teacher at Sky Vista Middle School in Aurora, Colorado.

References

Bryson, B. 2004. *A short history of nearly everything*. New York: Broadway.

Hakim, J. 2004. *The story of science: Aristotle leads the way*. Arlington, VA: NSTA Press.

Hakim, J. 2005. *The story of science: Newton at the center*. Arlington, VA: NSTA Press.

Hakim, J. 2007. *The story of science: Einstein adds a new dimension*. Arlington, VA: NSTA Press.

FIGURE 6

Concept for consideration: How does the study of seismic waves and the discovery of convection currents in the mantle help our understanding of continental drift and plate tectonics?

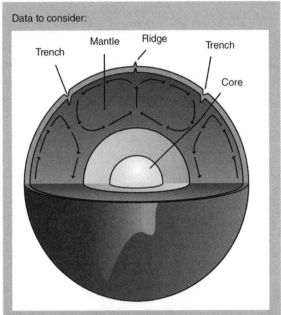

Data to consider:

By studying how earthquake shockwaves travel through the Earth, a discovery has been made about the layer of the Earth's interior called the *mantle*. It appears that the mantle is made of rock that has been heated so that it is not truly solid, but somewhat softer and can flow. As the rock near the core gets heated to extreme levels, the material starts to rise. As the material reaches the underside of the crust, it spreads out, cools down, and begins to sink. This is referred to as a *convection current*. There have been a number of these convection current cells discovered, as are shown in the image above. What implications might this have for the Earth's crust?

This article is reprinted with permission from Cronin, J. 2008. The 23rd Annual Consortium of Geologists. Science Scope 32 (2): 25–29. The 23rd Annual Consortium of Geologists. By Jim Cronin.

APPENDIX 5

Model Science and Math Activity (K–5)

Activity 10: Investigating the Properties of Magnets

Overview

In this activity, younger students encounter, discuss, and apply the basic characteristics of magnets and magnetism as they explore and elaborate on their experiences. Student groups implement some of the terminology and concepts appropriate to the study of magnets as they investigate and measure how far magnets can repel one another and how many paper clips their magnets can attract.

Processes/Skills

- Observing
- Connecting
- Describing
- Analyzing
- Concluding
- Measuring
- Calculating
- Problem solving
- Communicating
- Cooperating

Recommended For

Grades K–4: Small group instruction

Offer students in grades K–1 extra assistance during the exploration in Procedure 1, and simplify the introduced vocabulary as necessary. You can also simplify the quantitative "repel" and "attract" investigations in Procedure 3 or even conduct those activities as teacher-led demonstrations.

Time Required

1–2 hours

Materials Required for Main Activity

- An assortment of shapes and sizes of small magnets (available through science supply catalogues, electronics stores, toy stores, and dollar stores)
- An assortment of miscellaneous objects and materials (such as paper clips, index cards, plastic pen caps)
- String
- Paper clips
- Metersticks or metric rulers

Connecting to the Standards

NSES
Grades K–4 Content Standards:

Standard A: Science as Inquiry

- Abilities necessary to do scientific inquiry (especially making good observations, planning and conducting a simple investigation, and communicating their ideas)

- Understanding about scientific inquiry (especially developing explanations using good evidence)

Standard B: Physical Science

- Light, heat, electricity, and magnetism (especially that magnets attract and repel one another and other materials)

NCTM
Standards for Grades PreK–2, 3–5:

- Measurement (especially understanding units and processes of measurement and measurable aspects of objects)

- Problem Solving (especially constructing new math knowledge through problem solving)

- Reasoning and Proof (especially developing, selecting, and evaluating arguments and proofs)

Safety Considerations

Basic classroom safety practices apply. Caution students that opposing magnets must not become flying projectiles as the magnets repel one another.

Batteries used to make electromagnets (in Other Options and Extensions) should be used only under adult supervision.

Activity Objectives

In the following activity, students

- actively explore magnets and their interactions; and

- determine, via quantitative and qualitative inquiry, whether magnetism is cumulative.

Main Activity, Step-by-Step Procedures

1. Begin the lesson with an open exploration of the magnets. Present student groups with some concrete materials (several magnets of varying shapes and sizes and a few miscellaneous objects such as paper clips, index cards, plastic pen caps), little or no introduction or direction by the teacher, and ample time to "play" with the materials.

    Simply give the groups time to do some trial-and-error exploration with the materials, under the condition that when the agreed upon time is up (15 minutes should be sufficient), each group will briefly report to the class regarding something that they learned about the materials and how they interact. (One student should be chosen to keep brief notes and drawings of the findings.) During the exploration time, circulate among the groups, gently facilitating an inquiry of the materials' interactions but not pressing students in any particular direction; encourage divergent and innovative thinking.

2. When the exploration time is up, allow each group to choose a spokesperson who reports one aspect of what the group members learned to the class (usually the person who took the notes and made drawings of the findings). As students talk about their findings, reinforce or introduce appropriate terms and concepts through nurturing and inquisitive dialogue. Depending on the grade level, such terms might include, but would not be limited to, *push, pull,*

attraction, repulsion, poles, attract, and *repel.* Concepts and terms may be reinforced or introduced through comments and questions. For example, "I notice that you said the magnets 'stuck together'—in science we might say that they were *attracted* to each other." Or, "You mentioned that the magnets could push away from each other, even through the index card. That's a very good observation. We could say that the magnets push or *repel* each other. Can the magnets do this no matter how they are arranged?" The point is for students to have an opportunity to use authentic dialogue when connecting science ideas to their own exploratory experiences.

3. Student groups can elaborate on their understanding by measuring how far one magnet can push another magnet of similar size and shape (instruct students to hold the magnets together and let one "spring" away in repulsion from the first). After predicting the number, students conduct five trials and record the data in Activity Sheet 10.1, Table 10.1 ([p. 235]; for older students, average the trials). Next, students measure and record how far two magnets can push a single magnet (Table 10.2 [p. 235]). Then students try it with three magnets pushing a single magnet (Table 10.3 [p. 236]). Graph the class results (number of magnets pushing versus distance pushed). Compare results and reach conclusions together as a class. Ask students, "Is 'magnet power' cumulative?"

4. Next, have each group tie or tape a magnet to a string and tape the other end of the string to the edge of a desk so that the magnet hangs freely in the air. Students then predict how many paper clips the magnet can attract and hold (Table 10.4 [p, 236]). To determine exactly how many paper clips the magnet can hold, students run three trials, record and average the data, and reach conclusions. After students are done with their trials, ask them to place the paper clips end-to-end so that the clips are just touching, but not hooked together (see Figure 10.1). Ask students, "How many paper clips can the magnet hold now? How do you explain the fact that the paper clips can attract other paper clips? Can paper clips do this when they are not touching a magnet?" Then say, "You saw how many paper clips could be held up by a single magnet. Next, predict how many clips can be held by two magnets tied together." Another option is to have each student group discuss, invent, illustrate, and report to the class on (though

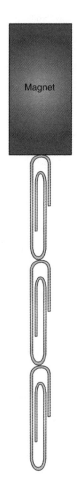

FIGURE 10.1.

Magnet and paper clips

Magnet

RUBRIC 10.1

Sample rubric using these assessment options

	Achievement Level		
	Developing 1	Proficient 2	Exemplary 3
Did students actively explore magnets and their interactions?	Marginally engaged in the exploration process	Satisfactorily engaged in the exploration process	Significantly engaged in the exploration process
Were students successful in connecting meaningful science terms and concepts to their exploratory experiences, and in applying those concepts and terms to their expanded investigations?	Some success with discussing terms and concepts but unable to connect or apply them to any significant extent	Satisfactory success with connecting the terms to their experiences and some success in applying them to their expanded investigations	Significant success connecting the terms to their experiences and in applying them to their expanded investigations
Were students successful in connecting meaningful science terms and concepts to their exploratory experiences, and in applying those concepts and terms to their expanded investigations?	Attempted the investigations, but unable to clearly determine	Able to clearly determine	Able to clearly determine and applied that understanding in discussions relating to their earlier explorations and findings with magnets

not necessarily build) an invention that uses magnets and/or magnetism.

5. Finally, find out what else the students want to know about magnets. Especially for older students, allow them to measure something else about magnets—something that they devise and want to know about, such as how much mass a particular magnet can lift or whether differently shaped magnets can lift different amounts of material. The groups should carry out and eventually report on their investigations.

Discussion Questions

Ask students the following:

1. What have you learned about magnets and magnetism? What did you do to find this out?

2. What do you still want to know about magnets and magnetism? How could you find out what you want to know?

3. How can magnets and magnetism be useful to us in everyday life?

4. When we measured how far one magnet would "spring away" from the other (Procedure 3), why did we conduct five trials and then calculate an average rather than just doing a single trial?

Assessment Suggestions for specific ways to assess student understanding are provided in parentheses.

1. Did students actively explore magnets and their interactions? (Use your observations of student activity during Procedure 1 as an embedded assessment.)

2. Were students successful in connecting meaningful science terms and concepts to their exploratory experiences, and in applying those concepts and terms to their expanded investigations? (Listen to student responses to Discussion Questions 1–3 during Procedures 1–5 as embedded evidence, or use the Discussion Questions as prompts for science journal entries.)

3. Were students able, via quantitative and qualitative inquiry, to determine whether magnetism is cumulative? (Use student responses to the exercises in Activity Sheet 10.1 [p. 235]as a form of performance assessment.)

Other Options and Extensions

1. Challenge students to figure out how an electromagnet works and ask them to make one from a battery and a length of insulated wire.

2. Ask students to figure out how an electric motor works. Ask them, "What do magnets have to do with it? Can you build an electric motor of your own?"

3. Ask students, "What does magnetism have to do with generating electricity? Can you find out how a generator works?"

Resources

Ashbrook, P. 2005. More than messing around with magnets. *Science and Children* 43 (2): 21–23.

Burns, J. C., and C. Buzzelli. 1992. An active attraction. *Science and Children* 30 (1): 20–22.

McCartney, R.W., S. Deroche, and D. Pontiff. 2008. Can trains really float? *Science and Children* 45 (7): 33–37.

Milson, J. L. 1990. Electromagnetic attraction. *Science and Children* 28 (1): 24–25.

Orozco, G. T., P. S. Alberu, and E. R. Haynes. 1994. The electromagnetic swing. *Science and Children* 31 (6): 20–21.

Sharp, J. 1996. Manipulatives for the metal chalkboard. *Teaching Children Mathematics* 2 (5): 280–281.

Teachworth, M. D. 1991. A memory for magnets. *Science and Children* 29 (2): 30–31.

Whitin, D. J., and P. Whitin. 2003. Talk counts: Discussing graphs with young children. *Teaching Children Mathematics* 10 (3): 142–149.

This chapter and the following worksheet are reprinted with permission from Eichinger, J. 2009. Activity 10: Investigating the properties of magnets. In Activities linking science with math, K–4. *pp. 87–93. Arlington, VA: NSTA Press. Figure numbers are from the original book.*

APPENDIX 5

Activity Sheet 10.1: Investigating the Properties of Magnets

1. How far can *one* magnet "push" another magnet (from Procedure 3)?

   (Prediction: _____ cm)

   Table 10.1

Trial	Distance (cm)
1	
2	
3	
4	
5	
Average	

2. How far can *two* magnets "push" another magnet (from Procedure 3)?

   (Prediction: _____ cm)

   Table 10.2

Trial	Distance (cm)
1	
2	
3	
4	
5	
Average	

3. How far can *three* magnets "push" another magnet (from Procedure 3)?

(Prediction: _____ cm)

Table 10.3

Trial	Distance (cm)
1	
2	
3	
4	
5	
Average	

Conclusions:

4. How many paper clips can the hanging magnet attract at one time (from Procedure 4)?

(Prediction: _____ clips)

Table 10.4

Trial	Distance (cm)
1	
2	
3	
Average	

Conclusions:

APPENDIX 6

Model Science and Math Activity (6–12)

Activity 14: Heat Exchange in Air, Water, and Soil

Overview

The Earth is composed, at least at the surface, of soil/rock, water, and air. How do the heat exchange properties of these three very different substances compare, and what effect, if any, do they have on climate and weather? This activity represents a somewhat more advanced investigation of a complex subject, but is readily adapted to lower grades.

Processes/Skills

- Observing
- Inquiring
- Describing
- Counting
- Measuring
- Graphing
- Analyzing data
- Problem solving
- Concluding
- Applying conclusions

Communicating Recommended For

Grades 6–8: Small-group or whole-class instruction

You can adapt the activity for grade 6 by conducting the investigation as a whole-class exercise, rather than in small groups.

Time Required

1–2 hours

Materials Required for Main Activity

- Plastic cups (large size)
- Alcohol-filled thermometers or temperature probes for computer or graphing calculator
- Soil (or sand)
- Water
- Desk lamp(s)
- Graph paper

Connecting to the Standards

NSES
Grade 5–8 Content Standards:

Standard A: Science as Inquiry

- Abilities necessary to do scientific inquiry (especially observing carefully, thinking critically about evidence to develop and communicate good explanations, and using mathematics effectively)

- Understanding about scientific inquiry (especially emphasizing the value of evidence and mathematics)

- Standard D: Earth Science

- Structure of the Earth system (especially regarding the properties of soil, water, and air)

- Transfer of energy (especially that heat moves in predictable ways)

NCTM
Standards for Grades 3–8:

- Measurement (especially understanding and applying the metric system)

- Data Analysis and Probability (especially displaying relevant data to answer questions)

- Reasoning and Proof (especially engaging in thinking and reasoning)

Connections (especially recognizing the connections among mathematical ideas and to investigations outside mathematics)

Safety Considerations

Basic classroom safety practices apply. If you use alcohol-filled thermometers (do not use mercury-filled thermometers), be sure to choose those with metal safety backs.

Activity Objectives

In this activity, students

- explore the heat exchange properties of the three substances (air, soil, water);

- make thoughtful connections between heat exchange and climate/weather patterns; and

- collect, display, and analyze their numerical data regarding heat exchange.

Main Activity, Step-by-Step Procedures

1. Ask students to recall a time when they sat outside in the sunlight—maybe at the beach or at a local park. Better yet, if it is in fact a sunny day, go outside and let everyone stand in the sunlight for a few minutes. Ask students how they would describe their experiences, especially in terms of temperature changes. They will probably say that they became warm. This is an example of *heat exchange,* in particular *heat conduction,* which is when heat flows from one object or substance (in this case, the air) to another (in this case, the students themselves). What do the students do when they become too warm? They leave the sunlight and somehow find shelter. But the Sun shines 24 hours a day, 7 days a week, and the Earth itself cannot find shelter from the Sun's light and warmth. Explore the following questions with the class: How do the substances of the Earth (air, soil, water) react to the warmth of the Sun? Are they all the same, or do they react differently? Do they differ in how they conduct heat? Do some of the Earth's substances retain heat better than others (that is, do some substances lose, or *radiate,* heat more slowly than others)? How could we find out? Encourage diverse responses

and try the ideas. One particular means of inquiry follows.

2. Divide the class into three groups (or multiples of three), one for each of the three substances mentioned. Provide each group with a large plastic cup filled with a single substance, that is, either a cup of air, soil (or sand), or water. Each group also receives a thermometer, which is placed into the central portion of their substances. Students should predict which substances will heat fastest and which will lose heat the fastest. That is, which substances will conduct and radiate heat most quickly? Students should be sure that all substances are initially at room temperature, and record that temperature. Students immediately place each of the three cups directly under identical desk lamps and record each substance's temperature every minute for 15 minutes, using Activity Sheet 14.1, [p. 243], to record data. At the end of that 15-minute period, students turn off the lamps and continue recording temperature readings each minute for the next 15 minutes. Results will vary significantly depending on how deeply the thermometer is inserted into the soil/sand and on how close the three substances are to the lamp, so all groups should try to be consistent.

3. All groups then share data so that everyone has results for all three of the substances. In the last column of the data table (Activity Sheet 14.1 [p. 243]) students can calculate the heat retained, if any, by each of the substances. (Find the difference between Minute 15 Temperature and Minute 30 Temperature.) Ask students how else they could decide which substance retains heat best, worst, and so on. Suggest they plot temperature (y axis) versus time (x axis) for all three substances on a single graph so that they can be easily compared throughout the 30-minute time period. (Each of the three resulting lines should be labeled so that students can tell them apart.)

4. Ask students, "According to the graph, which substance warmed the fastest? Which cooled the quickest? How do you know? Which substance retained heat most effectively? That is, which lost its heat most slowly? How close were your predictions to the actual results?"

   For more advanced students, the inquiry can expand into actual rates of heat gain and loss in degrees per minute.

5. Ideally, water generally retains heat most efficiently, followed by soil (or sand), and air. Water retains heat well and acts like a heat reservoir: It heats slowly and cools slowly. Explain that heat is conducted, but cold is not. Air, like most gases, is a poor conductor of heat because it has few molecules per unit of volume with which to transfer heat energy. Air (and water) can transport heat effectively via *convection*, however. Convection refers to the transport of heat energy by the actual motion of the heated gas (or liquid). Currents caused by convection are easily seen in boiling water. Solid land (e.g., rock, clay, sand, and soil) conducts and radiates heat efficiently, allowing it to heat *and* cool quickly. Heat exchange in real systems is quite complex because it is influenced by many variables,

including mass, evaporation, and reflectivity of the substances being heated.

6. Explain that the interplay among the temperature differentials of atmosphere, water (especially oceans), and land (especially continents) has a huge impact on the climate, weather, and the general circulation of the Earth's atmosphere. For instance, in coastal areas during the day, the land tends to be warmer than the adjacent ocean, and the opposite is true at night. This effect tends to stabilize the region's temperature and often causes daytime onshore breezes and nighttime offshore breezes, which affect the level of comfort for residents. On the other hand, continental heartlands, far from the influence of sea or

RUBRIC 14.1

Sample rubric using these assessment options

	Achievement Level		
	Developing 1	Proficient 2	Exemplary 3
Did students successfully explore the heat exchange properties of the three substances (air, soil, water)?	Attempted to explore heat exchange but were not particularly successful	Successfully explored heat exchange	Successfully explored heat exchange and could explain their investigation in detail using appropriate terminology
Were students able to make thoughtful connections between heat exchange and climate/weather patterns?	Attempted to describe the connections but were not successful to any significant extent	Successfully described the basic connections between heat exchange and some climate/ weather patterns	Successfully described in considerable detail, using appropriate terminology, the connections between heat exchange and some climate/ weather patterns
Could students successfully collect, display, and analyze their numerical data regarding heat exchange?	Collected their data but were unsuccessful in displaying and analyzing it	Successfully collected, displayed, and analyzed their data	Successfully collected, displayed, and analyzed their data and could explain the process in detail using appropriate terminology

ocean, tend to be very hot in the summer and very cold in the winter.

Discussion Questions

Ask students the following:

1. How do the thermal properties of air, soil, and water affect climate and/or weather?

2. How did mathematics help you determine the differences in heat exchange of the three substances (air, soil, water)? Could you have reached a meaningful conclusion without using math?

3. Why do some people want to live near an ocean or sea, in terms of the water source's effect on weather/climate? Would you like to live in this type of climate? Why or why not?

Assessment

Suggestions for specific ways to assess student understanding are provided in parentheses.

4. Did students successfully explore the heat exchange properties of the three substances (air, soil, water)? (Use observations made during Procedure 2 as a performance assessment, and use Discussion Question 1 as an embedded evidence or as a writing prompt for a science journal entry.)

5. Were students able to make thoughtful connections between heat exchange and climate and/or weather patterns? (Use Discussion Question 3 as an embedded assessment or as a writing prompt for a science journal entry.)

6. Could students successfully collect, display, and analyze their numerical data regarding

heat exchange? (Use observations made during Procedure 3 and Activity Sheet 14.1 as performance assessments, and use Discussion Question 2 as an embedded assessment or as a writing prompt for a science journal entry.)

Other Options and Extensions

1. Have students design and build model houses of different substances (clay, cardboard, paper, etc.). Instruct them to test the heat retention of the houses, using a lamp as a heat source. Students should record the temperature data and find out which home material insulates best. Students also can test different designs for thermal differences.

2. Give each student group an ice cube (in the bottom of a clear plastic cup so that it is visible and you can see when it has melted) and challenge them to insulate the cube so that it lasts as long as possible under a desk lamp. Offer a variety of materials (cotton, paper, cardboard, packing bubbles, etc.) to place over the cube for insulation against the heat of the lamp.

3. Have students test the thermal properties of saltwater (using the basic procedure in Step 2), and compare the thermal properties of saltwater to those of freshwater. Students can compare the thermal properties of different types of soil and/or compare moist air (putting a damp paper towel in the bottom of the cup) with dry air for heat gain and loss.

Resources

Buczynski, S. 2006. What's hot? What's not? *Science and Children* 44 (2): 25–29.

Cavallo, A. M. I. 2001. Convection connections. *Science and Children* 38 (8): 20–25.

Chick, L., A. S. Holmes, N. McClymonds, S. Musick, P. Reynolds, and P. Shultz. 2008. Weather or not. *Teaching Children Mathematics* 14 (8): 464–465.

Critchfield, H. J. 1983. *General climatology.* Englewood Cliffs, NJ: Prentice-Hall.

Gates, D. M. 1972. *Man and his environment: Climate.* New York: Harper and Row.

Damonte, K. 2005. Heating up, cooling down. *Science and Children* 42 (8): 47–48.

Pearlman, S., and K. Pericak-Spector. 1995. Graph that data. *Science and Children* 32 (4): 35–37.

Whitin, D. J., and P. Whitin. 2003. Talk counts: Discussing graphs with young children. *Teaching Children Mathematics* 10 (3): 142–149.

This chapter and the following worksheet are reprinted with permission from Eichinger, J. 2009. Activity 14: Heat exchange in air, water, and soil. In Activities linking science with math, 5–8. 141–147. Arlington, VA: NSTA Press. Figure numbers are from the original book.

APPENDIX 6

Activity Sheet 14.1: Heat Exchange in Air, Water, and Soil

Predict: Rank order the three substances (air, soil, water) in terms of how well you think they will be able to retain heat.

Ranking	Prediction	Actual
Most Effective		
Second Most Effective		
Least Effective		

Record heat gain (conduction) and heat loss (radiation) data for all three substances. Each group will experiment with a single substance.

APPENDIX 6
Activity Sheet 14.1: Heat Exchange in Air, Water, and Soil

Room temperature (initial temperature of all three substances) = _____ °C

Data Table

	Lamp on (First 15 Minutes; Record Temperature)															Lamp off (Last 15 Minutes; Record Temperature)															HR*
	1	2	3	4	5	6	7	8	9	10	11	12	13	14	15	16	17	18	19	20	21	22	23	24	25	26	27	28	29	30	
Air																															
Soil																															
Water																															

* HR (Heat Retained): Minute 15 Temperature minus Minute 30 Temperature

Graph the class data for all three substances: time (*x* axis) versus temperature (*y* axis). What can you conclude based on the data?

INDEX

*Page numbers printed in **boldface** type refer to tables or figures.*